SECRET WORLDS

the **extraordinary** senses of **animals**

martin
STEVENS

SECRET
WORLDS

OXFORD
UNIVERSITY PRESS

OXFORD
UNIVERSITY PRESS

Great Clarendon Street, Oxford, OX2 6DP,
United Kingdom

Oxford University Press is a department of the University of Oxford.
It furthers the University's objective of excellence in research, scholarship,
and education by publishing worldwide. Oxford is a registered trade mark of
Oxford University Press in the UK and in certain other countries

First Edition published in 2021

Impression: 2

Published in the United States of America by Oxford University Press
198 Madison Avenue, New York, NY 10016, United States of America

British Library Cataloguing in Publication Data
Data available

Library of Congress Control Number: 2021934078

ISBN 978–0–19–881367–5

Printed and bound by
CPI Group (UK) Ltd, Croydon, CR0 4YY

For Lenny, Sam, and Audrey

PREFACE

The philosopher Thomas Nagel once posed the question 'What is it like to be a bat?' in a thought experiment about perception and consciousness. In doing so, he was really asking questions about what consciousness means, but we might equally pose the same question to think about what the sensory world of a bat is like and how that governs its life. It is easy to fall into the trap of assuming that other animals perceive the world in the same way that we do, but the reality could hardly be more different, or exciting. What we perceive are but snapshots of the physical world, measured and interpreted by our senses and our brain. The product of our evolutionary past, our senses only allow us to perceive those aspects of the world for which we have the necessary apparatus, used to gather the information we needed to survive.

Far from being uniform across species, how an animal perceives the world is heavily dependent on its sensory systems and brain. In the case of bats, many species have a highly sophisticated ability to use echolocation to navigate and hunt for prey. This is centred on ultrasonic frequencies that operate well above our hearing range. Our ears are simply not tuned to detect these frequencies. Each animal's perception of the world is therefore a product of its sensory systems, and the information detected can differ greatly from other species.

It's worth pausing for a moment to consider how crucial our senses are to everything we do. Vision, smell, taste, hearing, and touch all provide us with an essential ability to respond to threats, communicate with one another, perform numerous daily tasks, avoid obstacles, and interact appropriately with the world around us. To people who have lost just one of their main senses, such as vision, many tasks the rest of us take for granted can be challenging. So, imagine what life would be like if we lost our vision, our hearing, our smell and taste, our touch, and so on. Our senses provide a critical gateway to the outside world, allowing us to interact with it. The same is true for all animals—their sensory systems are what enables them to forage, avoid predators, attract mates, navigate, and much more. Without them, individuals would be completely helpless.

Human senses do a reasonable job of allowing us to gather information from the world and behave accordingly. But throughout this book we will encounter many animals with senses that, in comparison to ours, seem extremely refined. By contrast, we are more of a jack of many trades, with a range of good but not spectacular senses. I say 'many' rather than 'all' trades because we lack entirely some sensory systems that other animals possess. Ultimately, the sensory systems of other animals are tuned to widely different stimuli. For example, many animals, from spiders to birds, can detect and respond to ultraviolet light, to which we are blind. Others, including numerous insects, rodents, and bats, can hear high-frequency ultrasonic sounds well beyond our own hearing range. And this is just the tip of the iceberg. Consider the magnetic sense of birds, turtles, and other animals, or the electric sense of many fish and some mammals. Such great differences in sensory ability reflect adaptations to different habitats and lifestyles. And, when sensory systems adapt to different parts of an environment and affect behaviours such as mate choice, they can even drive the formation of new species.

This book is about the remarkable sensory worlds that animals experience, often so different from our own. It will explore how different animal senses work; what they are used for; how they evolved and were shaped by the ecology of a species, the environment where it lives, and the tasks it must complete during its life. Throughout, we will also see how scientists investigate the ways that animals use their senses, given that we can't experience their sensory worlds at first hand. What I hope to show is how such work reveals the remarkable diversity of animal life, and how the study of sensory systems has shed light on some of the most important aspects of behaviour and evolution.

Work on animal senses covers vast ground and goes back a very long way. Some of the earliest evolutionary biologists, not least Charles Darwin and Alfred Russel Wallace in the mid-to-late 1800s, and many philosophers well before them, investigated how the senses function and posed questions about how they differ among species. Much has been learned over decades of research, but there are many outstanding questions too—such as how the magnetic sense actually works at all. It was a challenge from the outset to decide how to frame this book and what to cover. From the start, I decided simply not to cover everything—there is far too much. Instead, for each of the main chapters, I consider one sense in turn, focusing primarily on three animals or animal groups. Of these, I have picked examples of species that we understand quite well, or which are particularly remarkable in how their senses operate or in their level of specialization, although I touch on other examples too for wider context. Together, these three animals per chapter tell the story of how different senses work, and also illustrate broader issues regarding the role of animal senses in the ecology and evolution of all species. I have been a little loose with the three-animal rule when needed—sometimes one very specific species forms an adequate example, when it has been the focus of a concerted research agenda. This is the case for the star-nosed mole and its sense of touch in Chapter 6. On other occasions, much of what we know comes from a wide range of similar species, as is the case with groups of fish that all produce and detect low levels of electricity. But the idea is the same: each animal or group reveals the wonderful nature of animal senses, how they operate, what they are used for, and their importance to an animal in the environment in which it lives.

The original idea for this book has a fairly long history. Back in 2013, I published a textbook, also with OUP, on sensory ecology. This was never intended as light reading, but rather as a reference book for students and scientists. Still, it was impossible not to think about how exciting the subject would be to wider audiences. I then wrote a book for a wider readership about deception in nature (*Cheats and Deceits*, 2016). This turned out to be a very enjoyable experience, and it was almost immediately obvious that I should write a similar one on animal senses. This is it.

Many people have helped with this book and supported me during the process. My wife Audrey has, as always, been wonderfully supportive, encouraging, and enthusiastic throughout, and my two boys Sam and Lenny have provided endless inspiration with their own fascination for the natural world. I am very grateful to an anonymous reader for a range of helpful feedback, and Jim Galloway and Katya Zaki for their comments on an early draft. Particular thanks are due to my editor Latha Menon for an abundance of feedback, guidance, and advice on the book and the process as a whole. Her various inputs and critiques have been invaluable. I also thank Jenny Nugee at OUP for a wide range of help and input on the book and its contents. Finally, my research group past and present, and the various undergraduate students who have taken my sensory ecology lectures, first at Cambridge and then Exeter, have provided an endless source of inspiration and enthusiasm for the subject.

CONTENTS

A PLETHORA
OF SENSES

The Caribbean spiny lobster (*Panulirus argus*) lives on coral reefs in the western Atlantic. In the autumn, you can witness thousands of these lobsters all migrating en masse. It makes an incredible sight. The creatures head out to sea and away from the shallow coastal waters, often in head-to-tail queues occurring over both day and night. The procession is remarkable both in numbers but also in formation, with each lobster holding out its antennae to touch the back of the individual in front, an arrangement that reduces drag in the water. Their movements are driven by the onset of the first autumn storms in shallow seas, which stir up the water and lower its temperature.

During these events, the lobsters in a particular area all tend to walk in the same direction, as if guided by some invisible force. They head some 30–50 km away from the shore and down to depths of up to 30 m, where the waters are calmer. Outside of the migratory period, when they are still living on the shallower reefs, the lobsters lurk hidden away during the day in one of a number of dens, only emerging at night to feed. Once they venture out from the safety of their shelters, each individual wanders over a wide area. As the morning arrives, they return in a straight line back to their homes (Figure 1). Clearly, the lobsters have some sort of ability to determine where they need to head, both when migrating and in order to find their way back to their dens every morning. How do they do this?

In the 1990s, the biologists Kenneth Lohmann and Larry Boles from the University of North Carolina, together with colleagues, set out to find the answer.[1] They suspected that the lobsters are governed by a mysterious magnetic sense. The team captured lobsters on the Florida Keys by diving and prodding them out of their crevices with rods, or 'ticklesticks'. They then covered the eyes of the lobsters with eye caps, to prevent them from using visual information to guide their movements, and tethered them to underwater walking areas that were surrounded by a magnetic coil system. In this way, the scientists could alter features of the

Figure 1. A line of Caribbean spiny lobsters migrating across the Atlantic Ocean floor during their autumn movements into deeper waters.

Earth's magnetic field around the lobsters. And, sure enough, the experiments began to uncover the hidden sensory skills of these animals. The team found that the lobsters tended to move in certain directions, but when they changed components of the magnetic field, after a few minutes the animals walked in different orientations. It looked as though the lobsters had some sort of internal compass that could guide where they should go.

Some years later, the same team tried another kind of experiment. They moved a sample of lobsters a long way from their original homes, between 12 and 37 km away, in fact. In the process, some lobsters were also prevented from gathering information about the Earth's magnetic field (or visual cues), by using magnets hung inside the transport vehicles. This meant the displaced lobsters would have to navigate back to their homes by working out their new magnetic position relative to where their home should be. The lobsters were able to do just that; they could navigate home. They were probably able to do this by using some sort of internal magnetic map of their surroundings, because when the scientists put lobsters in magnetic fields that simulated areas corresponding to real-world locations, the lobsters moved in the correct directions that would get them home. If a lobster was presented with a magnetic location north of its territory, it would orientate south, and vice versa. These simulated locations were large distances (around 400 km) away from the lobsters' home sites, but the lobsters are thought to range this far. By knowing the magnetic features of their surroundings, these creatures can work out their position and then return to a specific spot. This goes beyond how we would use a compass to orientate; it's more like having an in-built GPS. Quite how the magnetic sense works in lobsters and many other species is a mystery of Sherlock Holmes proportions, and we'll come back to it in Chapter 7.

Other animals use different yet equally highly refined senses to survive and reproduce. The common vampire bat (*Desmodus rotundus*), from Central and South America, seeks out mammals, perhaps a wild tapir or domesticated cattle, in order to feed on their blood at night. Many bats have elaborate structures on their face and around their noses that are used for enhancing their echolocation ability by helping to emit sound. The vampire bat instead has a nose structure that is packed with infrared thermal receptors. Unlike most other bats, vampires don't have to capture prey on the wing, but rather land near their prey and then scuttle on the ground towards them (presumably so they don't wake their victim by crash landing on them). Once within touching distance, the bat uses its infrared receptors not just to find suitable skin but to pick out blood vessels to bite into with its sharp teeth.

Detecting tiny differences in heat requires the evolution of a precision gauge operating in the right temperature range. The bats can do this because a gene known as *Trpv1*, present in other mammals and normally used for detecting high temperatures that might cause damage, has in vampire bats become modified by natural selection. Somewhere in the bat's evolutionary history, mutations occurred in the gene that altered the temperatures the receptors detect, and this was advantageous to those blood-feeding bats, helping them to find a meal more efficiently. Instead of preventing the animal from getting burnt, the modified gene now enables the sensory cells around the nose to detect much lower temperatures. The sensitivity of these cells has shifted into the temperature range for detecting mammalian body heat.[2] Such genetic changes, resulting in a slightly different role, happen often in the evolution of the senses, and in vampire bats it allows great refinement in their unusual habits.

Infrared and magnetic detection are senses that we humans lack; we have no conception of what it must be like to sense the magnetic field of the Earth and use it in any way, at least not without technology. The same is true for other non-human senses, not least the exquisite electric sense that a variety of fish and a handful of other animals possess. Even when it comes to the senses we do have, our abilities are restricted. We can't hear the ultrasonic calls used by many bats and insects, for example. Nor can we smell and interpret the changes in odour plumes blown around by the wind, originating from female moths and used by males to locate a mate from as far away as 10 km. In short, our senses are limited, and we perceive only a part of the world that is available to other animals.

It's natural to wonder what the world must be like to these species that perceive things that we cannot, but we should also ask why the senses vary so much among species anyway. These two questions are central to this book, but let's begin by considering the variety of animal senses; how they work; what they are used for; and why such a staggering array of senses exists in nature. As we will discover, the senses found across animal species, and even among individuals of the same species, vary according to many factors, not least the habitat in which an animal lives, the key behaviours it must perform in its life, and the costs and limitations of having and maintaining different sensory systems.

Animal senses are carefully refined through evolution and development for the things that matter most to them. To accomplish the numerous tasks every individual must perform, the senses are tuned to work best in the habitats where the creature lives and to acquire the best available sources of information. Sometimes, one or more of the senses are exquisitely tuned to a few crucial tasks an animal must perform in order to survive and reproduce successfully.

The parasitoid flies demonstrate this rather nicely, albeit gruesomely. The hearing organs in some of these flies have evolved extreme levels of specialization. Many of them find their hosts—male crickets—by eavesdropping on the calls that the males make to females during mating. Male crickets compete with one another by making high-pitched calls during courtship. It's a lovely sound on a summer evening to hear groups of male crickets chirping away, all in the hope of an amorous liaison. Quite often, females prefer males with the loudest or most elaborate calls as a suitable partner. Such males might be the best around with whom to mate and sire the fittest offspring. Unfortunately for the crickets, the parasitoids reproduce in a rather grisly way. Once they have located a suitable chirping host, they lay an egg on the male's back. When it hatches, the maggot burrows into the cricket's body, eating him from the inside out. To be successful, the parasitoid flies must locate males, and they can be very good at this: in some bush cricket populations, up to 60 per cent of the males are infested.

Although they may use their ears for other tasks, clearly both the fly and female crickets need to detect male calls. Take the example of the fly *Ormia ochracea*, which targets a specific field cricket, *Gryllus rubens*. Normally, these two groups of insects (crickets and flies) would have quite different hearing organs. However, in this case, not only are the ears of both the female cricket and fly finely tuned to detect the peak frequency of male calls (4–5 kHz), but the fly's hearing organ has a physical structure very similar to that of the female cricket.[3] The parasitoid's hearing has evolved along the same path as that of the female cricket, as both depend on detecting the call of the male cricket. This phenomenon, in which animal species or groups that have been exposed to similar selection pressures independently evolve similar adaptations, is known as convergent evolution. For example, dolphins share similar features of a streamlined body shape with certain prehistoric ichthyosaurs because it helps them move efficiently through the water. In the parasitoids, a task of great importance for the fly and for the female cricket has led to matching specialist tuning of a critical sensory system in both. As for the male crickets, sometimes the best males that call the loudest pay the cost of also being those most likely to fall victim to the flies.

Quirky, and somewhat less grisly, examples of sensory systems used to find prey abound in nature. One is the sense of smell (olfaction) of the 'vampire' jumping spider (*Evarcha culicivora*) found in East Africa. This spider feeds on the blood of vertebrates like humans, but it does so in a roundabout way. The spider hunts and kills female *Anopheles* mosquitoes (the type that carry human malaria), and they do this especially when the mosquitoes have fed on mammalian blood. Like many other jumping spiders, the vampire spider has excellent eyesight and can recognize mosquitoes based on their visual appearance. But more unusually for jumping spiders, it also has an excellent sense of smell which it uses to find its prey. Interestingly, the spider is known to frequent locations where humans reside. Scientists have investigated how these spiders in Kenya find mosquitoes, and why they are so often found with people, including how the spiders respond to a very specific stimulus—worn socks.[4] The researchers obtained cotton socks from a human donor who had worn them for 12 hours immediately prior to the experiments, and compared how the spiders responded to the smelly socks versus an identical pair of clean socks. The spiders were more attracted to the socks smelling of human than to the clean socks. So, the spiders seem to find humans based on their smell, in order to be in the right location where a key prey source is likely to visit.

Eyes also have many adaptations in nature linked to how animals go about their lives. The four-eyed fish (*Anableps anableps*) lives in northern parts of South America, inhabiting fresh and brackish waters. It floats at the top of the water looking for both predators and prey (mostly insects that fall into the water) from above and below at the same time. Half of each eye sits above the water, and half below. This creates a problem, because light is bent or refracted as it passes the boundary between air and water, and vision benefits from the eyes being able to form a sharp image. Our eyes have evolved for vision out of water and so things are blurry when we look underwater. Water, and any particles in it, also affects the wavelengths of light that are available. In clear water, longer 'red' wavelengths of light get filtered out earlier, shifting the light spectrum (the range of 'colours') towards blue and green light. Anyone who regularly dives in clear ocean waters knows that it becomes bluer as you descend deeper. Any particles in the water can absorb and scatter light, further changing the spectrum. Many freshwater lakes and streams look brown or green due to the scattering or absorption of light by the organic material floating around. The four-eyed fish has sophisticated tricks to deal with these challenges—it has divided its eyes into two, with two pupils, and two sets of photoreceptors, each of which focuses and analyses the spectrum of light coming from either above or below the water line. It's probably about the closest thing we have on Earth to a vertebrate with four eyes. In doing so, the fish can see the world in sharp focus both above and below the water at the same time, and utilize differences in sensitivity in the two parts of the eyes to best see in the different light conditions. Quite how the fish's brain is able to process these two views of the world is another matter waiting to be investigated.

Almost certainly the most important factor that dictates what senses an animal has, and how they work, is the environment in which the creature lives. The Mexican tetra (*Astyanax mexicanus*), also known as the 'blind cave fish', mostly live, as the name suggests, in dark caves, so their vision is of no use, and their sight has degenerated. Instead, they invest in other senses, especially those that respond to vibrations and water movement. Other animals that spend their entire lives underground in caves, so-called troglobites, likewise commonly have no vision but instead have elongated antennae, sensory hairs, and various other adaptations for sensing chemical and tactile information. These underground beasts are often of bizarre appearance, being ghostly pale since they tend to lack pigments in their bodies. After all, there's no need for colour or pattern if there is no light for anything to see them.

While there are many species of troglobites, lifestyles need not always be so extreme. Many species spend only part of their lives in caves, or they simply live in habitats where light and visibility are restricted, or they are primarily nocturnal. Such animals also tend to invest more in senses other than vision. For example, electric fish, which are capable of producing weak electric fields for navigation and communication, tend to be nocturnal or to live in water that has poor visibility. Likewise, bats with their remarkable sense of echolocation tend to hunt at night and live in caves and crevices where vision is of less value. That said, things sometimes go the other way. Species that are active at night, as long as there is still enough light to see, may instead have very enlarged and sensitive eyes, as do some nocturnal primates such bush babies and birds like owls. But the general point is that the environment in which an animal lives, and when it tends to be active, dictates the senses in which evolution has tended to invest.

The environment is critical in tuning the way the senses work too, and in the specific information they collect. In the deep ocean, aside from being very dark, the cutting out of longer, redder wavelengths by the seawater results in light that is more blue-green. If there is no red light to see, then there's no point in having a visual system tuned to see longer wavelengths. And, sure enough, the majority of deep-sea animals have eyes that can detect shorter to medium wavelengths of light, but rarely red light. This also explains why many deep-sea animals are red: in the depths of the ocean, since there is no red light, they just look very dark and are well hidden. Only when we bring them to the surface and out of their natural environment do they take on a red colour to our own eyes.

The deep ocean is also rich in a variety of bioluminescent light, emitted by a plethora of creatures from jellyfish to squid. Far from being desolate and devoid of life, the deep-sea can be a constant fireworks display of flashes of light, from creatures which emit these bursts for a multitude of purposes: some to startle or scare predators; others to lure prey or attract mates. Again, this light is mostly blue-green in colour, so there is rarely a need for a visual system that can see red in the twilight zone. Indeed, many deep-sea fish, such as some shark species, have visual systems tuned to detect blue bioluminescence, either of prey or of one another.

On land, too, the times of day that animals are active also drive differences in vision. Among several species of *Myrmecia* ant from Australia, for example, the species and the worker castes among them that are more active at night tend to have larger eyes, and cells in their eyes that are more sensitive to low light levels.[5] I've focused here on vision to illustrate how an animal's environment influences its sensory systems, partly because animal vision has been well studied, but the same applies to other senses too, to a lesser or greater extent.

The environment can also place constraints and challenges on how the senses actually work. For instance, if there is a lot of noise, then an animal may need an auditory system that can overcome this. Some species of echolocating bats modify their calls to avoid interference from the noises made by other bats and insects that use similar frequency ranges. They do this by shifting the frequencies of their calls.

When it comes to detecting chemical information, we'll be looking here mostly at animals' sense of smell rather than that of taste. Nonetheless, the ability to detect chemicals through taste illustrates how the environment can limit how some senses work. Scientists have identified five basic types of taste thought to exist in vertebrate animals in nature (sweet, bitter, salty, sour, and umami). For some reason, most birds seem to have lost the ability to sense sweetness. Penguins also lost the genes encoding the receptors used for sensing umami and bitter tastes some 20 million years ago.[6] This seems odd because umami is thought to give rise to a sense of 'meaty' tastes, and penguins are carnivorous. The answer potentially lies in the environment where they live, and how the proteins that enable these sensations to work are affected by temperature. A key basis for adaptive evolutionary change is mutations, which can lead to alterations in some aspect of an animal's morphology or behaviour. Beneficial mutations that might improve features of an organism's biology, such as its sense of taste, can be subject to selection if the bearer of the mutation is better able to survive and reproduce. This would have been the case for the mutated gene *Trpv1* in the vampire bat, which, as we saw earlier, enables the bat to detect small changes in temperature, improving its efficiency in feeding. Conversely, mutations can also be deleterious, producing traits that are less useful or even damaging. Alterations of that nature will often be selected against, should they be sufficiently detrimental to fitness. Yet traits may also be degraded by mutations that render them less effective, but those changes might not be selected against if there is actually little impact on how an animal lives and reproduces in its current environment. Returning to penguin taste, a particular protein (*Trpm5*) seems very important in the sensory receptors that encode sweet, umami, and bitter sensations, but this does not work at cold temperatures. Penguins originated in the Antarctic around 60 million years ago, and diverged into different species approximately 35 million years later. The intervening time period coincides with phases of significant climate cooling in Antarctica, and penguins are still most often found in very cold conditions. Over the period of initial penguin evolution and climate cooling, penguins may have lost these types of taste through the build-up of mutations because their sense of taste simply does not work properly in the icy cold. In fact, penguins also seem to have very few taste buds. Maybe we shouldn't be too surprised by this, since we ourselves don't taste things quite so well when food is cold.

In many of the animals we will encounter in this book, one or two senses are supremely developed, like the hearing of a bat or the vision of a bird, but this is rarely the case for all their senses. Another approach is for species to adopt something of a jack of all trades, with a couple of preferred senses but a reasonable all-round set, even if none of them are particularly remarkable—a bit like humans perhaps. Not having amazing senses all round in part comes down to one key thing: the costs involved. We might not realize it, but our senses (and those of any animal) are expensive to run. For sensory and nerve cells to function, the body must constantly pump charged ions like sodium and potassium across the cell membranes so that the electric signals with which these cells communicate can be generated. This process consumes energy. If we take a fairly unglamorous animal, the blowfly (*Calliphora vicina*), scientists have calculated that around 10 per cent of all the energy it uses up while in a 'resting state' (that is, not flying around) is simply due to the functioning of the photoreceptors and associated nerve cells in its eyes.[7] In fact, blowflies have quite impressive vision. They use it to fly around rapidly and interact with objects (like other flies) at very short notice, so they need to invest a lot in their visual system. Nonetheless, similar calculations could no doubt be done for a range of animals, and if you were to add up the energy costs of not just the vision of an animal, but also its sense of hearing, taste, touch, smell, and any other senses it has, the costs would be pretty high.

If there is one thing evolution is good at, it's making the best use of the energy that animals have available to them. On the most fundamental level, sensory systems respond to changes happening in the environment. The reason for this is simple—when something changes it tells you something new, and that may require some action on your part, perhaps moving to avoid a threat or obstacle. But when things are the same, you need not worry. So, responding only to changes in the world around you is a sensible thing to do, and cuts some of the cost of continuously processing all the steady sensory information at the same time. Remarkable tricks appear over and over again in animals for encoding information that is specific to what they need to know, and at least two common ways to manage the costs of sensory systems emerge. One is to create nerve circuits and processing mechanisms that are efficient, and sometimes even focused on a specific task (such as detecting prey). We will come across many ingenious solutions throughout the book for how animals and their senses do this.

The other way to manage energy costs is to invest most in the senses that matter more to you, and either reduce investment in those senses that are less important, or give up on them entirely. Here, when we say 'invest', this is shorthand; what we mean is that there has been selection pressure that favoured individuals with new mutations that allowed them to expend more resources on certain traits over others. As we might expect, evolution is good at doing this, though the evidence has come from some surprising places. Fruit flies (*Drosophila*) are ubiquitous model species for studies of evolution, genetics, cellular biology, and much more in the lab. They are easy to keep, breed quickly, and an enormous amount is known about how their bodies work. Populations of fruit flies originating from wild populations have been kept in labs for decades, and most lab settings represent a rather unthreatening environment for them. The flies get food and water, they rarely have predators to worry about, and conditions are often stable and comfortable. The upshot of all this is that they do not particularly need all their senses to the same extent that their wild counterparts do. Flies in lab populations that have been in captivity for many generations have evolved smaller eyes.[8] In this instance, some flies have incurred mutations that caused a reduction in eye size, and led to less energy being invested in vision as a whole. Selection favoured flies with these mutations since they had little need for more advanced vision and could instead allocate the resources saved elsewhere. This is an example of natural selection, albeit arising in the artificial environment of a lab; the flies that invest less in vision save energy for other tasks and activities.

In the wild, there is also plenty of evidence for how evolution can change investment in the senses and phase out systems that are not needed. We have already come across the Mexican tetra or blind cave fish, which lives in parts of Central America and some neighbouring regions, mostly in caves where light doesn't penetrate. The surface-dwelling counterparts have functioning eyes, while the vision of the cave fish populations has degenerated to differing extents. Those that live entirely in caves have lost not only their vision but their eyes, whereas those living partially in the dark often have diminished visual abilities. Researchers have calculated that the eyes and the corresponding area of the brain used for processing visual information (optic tectum) use up about 15 per cent of the fish's energy costs in juveniles at rest, and about 5 per cent in adult fish.[9] These are substantial costs, so it is no surprise that fish living in dark caves have lost their eyesight; keeping up vision would be a huge waste of resources (Figure 2).

The example of the cave fish also illustrates how effectively tuned animal senses are to their environment, and the information available to them. Many species of fish have a sense that enables them to respond to changes in water movements and pressure—the so-called 'lateral line'. Up close, you can see parts of the lateral line as a row of dots running along the side of the

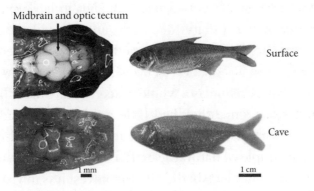

Figure 2. The Mexican tetra (*Astyanax mexicanus*) exists in populations of individuals that live in surface waters (upper image), and populations that are restricted to dark caves (blind cave fish). In the cave-dwelling individuals (lower image), the eyes have been lost and the area of the brain devoted to processing visual information is much reduced. This saves the fish wasting energy on a sensory system that is of no use.

body and around the head. This system is important for controlling movement, responding to currents in the water, detecting objects like prey, and responding to other fish. So refined is this sense in some fish that individuals can maintain their position and distance from others when shoaling. The lateral line is made up of special sensory cells called neuromast cells. The cave-dwelling populations of the tetra may have lost their eyes, but they respond very effectively to water vibrations, especially to obtain food that may fall into the water within the cave. Surface individuals on the other hand have functioning vision but relatively poor ability to detect vibrations. The cave dwellers' extra sensitivity to vibrations comes from having larger neuromast cells, as well as more sensory cells along their lateral line. Evolution is economical, allocating energy where it is of most value to an animal in a given environment.

The cost of sensory systems can be reduced in animals not just by evolution over many generations, but also during the lifetime of an individual. An extreme case can be seen in the aptly named ogre-faced spider (*Deinopis subrufa*) from Australia. This remarkable creature hunts at night, waiting for unsuspecting victims. It is also called the web-casting spider because it hunts by creating a net of silk, held between its legs, and casting the net over passing prey, ensnaring it. This requires excellent vision, especially in the low light levels when it is active. And, indeed, its forward-facing main eyes are enormous, taking up much of the face; it's an incredible-looking beast (Figure 3). The problem for the spider is that having such large and sensitive eyes is presumably very costly, and research back in the late 1970s showed that it does something very curious with its vision; it degrades and inactivates its sensory system when not in use.[10] Before night arrives and it starts hunting, the spider synthesizes new photoreceptor membranes in the eyes over about 1–2 hours. Then, at dawn, it

Figure 3. The ogre-faced or net-casting spider (*Deinopis subrufa*). This spider hunts at night by ensnaring prey in a web-net that it holds in its legs. Its huge eyes take up a lot of energy but enable it to see prey on the ground with high sensitivity.

breaks the membranes down again. This seems rather odd, but the likelihood is that it is energetically cheaper to break down parts of the visual system and remake the membranes each 24 hours than it is to maintain its top-quality vision during the daytime when it is not active.

Investment in different senses can also be flexible depending on the conditions under which an animal grows up. We know that through adaptation over many generations, species like blind cave fish can decrease investment in one sense (e.g. vision) in favour of another (e.g. mechanical), but individual animals can also do this during their lives. Guppy fish (*Poecilia reticulata*) are a beautifully coloured species found in places such as Trinidad and elsewhere in Central America, and widely kept as pets; virtually any tropical fish shop will have a tank of them. They live in streams, and males especially are brightly coloured with oranges, blues, blacks, and other colours, often shiny and iridescent. The ones seen in fish shops have been enhanced through artificial breeding, but the wild-type specimens are also wonderfully coloured, and, owing to their appearance, ease of keeping, and propensity to breed, widely used in scientific studies. Guppies use their vision to see one another and find food, but when individuals are reared under dark conditions, where vision would not be of great benefit, they invest more in their sense of smell and use this to locate things to eat. So sensory systems are far from fixed; they can change during an animal's life, and respond to the world in which they have developed. We'll come across this again, especially with regards to vision in Chapter 3. Maybe this capacity for change shouldn't come as too much of a surprise: it is often reported, albeit not always based on clear evidence, that people who lose one sense can compensate with improvements in another. Certainly, the idea of sensory systems and brain capacities remaining fixed during an animal's life is now largely an outdated concept.

In this book, I've taken the approach of looking at the different senses one by one. This is hopefully logical, not least because I focus in each chapter on examples of animals that have a remarkable sense in each modality. But there is a pitfall here too—it would be easy to fall into the trap of assuming that the senses work in isolation. In reality, information from several senses is often combined when performing specific behaviours. When choosing what to eat, we might smell foods as well as looking at them before picking what we want. In fact, one of the challenges in testing how animals use a particular sense is that, when scientists block or reduce the ability of an animal to use it, the creature may switch to relying on another sense to perform the task. The original focal sense may be the preferred one for the animal to use, but the animal could still be capable of employing other modalities when needed. This means that it can be hard to test how animals use their senses. So, before we consider animal senses one by one, let's look briefly at how the senses are combined.

One way to use multiple sensory systems is to rely on each one sequentially when performing a task. When sea turtles hatch and emerge from their nest on a beach they need to get to the ocean, and then into warm currents, and off to feeding grounds and relatively safer waters. Initially, when turtles hatch they find their way to water by using visual cues: they hatch at night, so they can utilize the fact that the light from the moon and stars is reflected more strongly from the water surface than from the land. They also often move downslope towards the water line. When in the water, they are guided by sensing the direction of the incoming waves. A hatchling's first priority is to head out to sea and away from the beach; and swimming directly against the waves is a good way of doing this, since the waves will be heading in towards the land. If the waves are coming towards the turtle from its right, the hatchling turns to the right and orientates into the waves, and the turtles can judge the circular motion of the waves to do this. Finally, the turtles use a magnetic sense to find and stay in key currents—until the time comes for females to return to the beach they came from years later so they too can nest, and then they do this using magnetic as well as probably odour and visual information. We'll come back to turtles and their use of a magnetic sense in Chapter 7.

Another, admittedly less endearing, example of how multiple senses are used is in how mosquitoes detect human hosts. The culprit in question is the *Aedes aegypti* mosquito. I have personally encountered these on many occasions while travelling in the tropics, partly as they tend to be active during the day, and they can be quite well marked with stripy colour patterns. Unfortunately, they are also vectors for a wide range of nasty diseases, including dengue fever, Zika virus, and yellow fever, to name but a few. Mosquitoes have an irritating habit of emerging from the depths of the jungle undergrowth just as someone walks by. Research has shown that they first detect our presence by chemically sensing the plume of carbon dioxide created when we exhale.[11] This then triggers a visual search stage, with the mosquito especially attracted to dark shapes and features around 5–15 m away. Once they get within around 20 cm of us, they use a thermal sense to establish where exactly to bite into blood vessels. It's an impressive sensory repertoire to deploy in order to find a meal.

Combining information from the senses simultaneously— something often referred to as sensory integration—gives even more power. This can greatly improve the accuracy of many behaviours. In Chapter 2, we will delve into the remarkable hearing of the barn owl (*Tyto alba*), which it uses to find prey in the darkness. Without spoiling that story, all we need to know here is that barn owls often combine their hearing with sight. When a barn owl hears a sound of potential interest, it very rapidly turns its head to orientate towards the source of the sound. Barn owls can locate the source using hearing alone, but while they move more quickly to sound cues, they tend to be more accurate when using visual cues—and so combining these senses makes the owls both fast and able to strike with pinpoint accuracy. However, this only works when information from these two senses arrives at the same time and is in agreement; if there is a conflict between the two senses then performance is actually made worse. We also know that there are specific nerve cells in the brain of the owl that respond to particular combinations of auditory and visual information that coincide in space and time, giving the owl a precise map of what is going on around it.

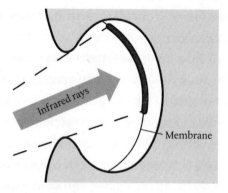

Figure 4. The pit organ of infrared-sensitive snakes can work in a broadly similar way to a pinhole camera or simple eye. Infrared rays enter the organ through a small opening and stimulate sensory cells on a membrane towards the back of the structure. The part of the membrane that is stimulated can provide information not just regarding heat but also on the size and direction of the stimulus (e.g. prey item).

Mosquitoes and vampire bats are two animals that have a thermal sense for detecting blood vessels. The pit-organ-bearing snakes are also masters of this sense. These vipers, pythons, and boas have small openings (you can see them by eye if you look closely) on their head, at the back of which is a thin membrane of receptor cells. Superficially, these 'pit-organs' are structured a little bit like an infrared version of a 'pinhole camera': thermal information from a particular source arrives at a specific angle through the cavity opening and onto an area of the membrane at the back, allowing the snake to deduce the direction to the source (Figure 4). Work on pit vipers shows that an important aspect of the sense is in being able to tell the difference in heat signatures between objects and the general environment. The cells in some snakes have been reported to be able to detect changes in temperature as low as 0.0003 °C, and while we should be somewhat cautious about such tiny and precise numbers, there is no doubt that the sense is extremely refined. One of the main functions of the infrared sense is that it enables the snakes to detect and attack prey, such as small rodents.

The question is how snakes actually detect suitable prey. This is not trivial because in the environment there will be a host of objects that will vary in temperature. To circumvent this problem, the snakes have some very clever processing tricks, combining infrared information with their vision to tell different objects apart. Scientists have studied the neural basis of infrared sensing by snakes since the 1930s; and how snakes combine this with vision has been studied since the 1970s and early 1980s. This work shows that the infrared sense encodes contrast in heat signatures against the background, and that the snakes form a combined map of infrared and visual information in their brain, in the optic tectum (a bit like the owls). The cells in the optic tectum respond to particular combinations of vision and infrared, and in doing so they effectively become 'feature detectors'. There are several types of cell that only fire when both the infrared and visual system are activated (AND cells), and these encode small, moving, warm objects: something like a mouse or other prey running around. Other cells are activated by warm but stationary objects (like a rock heated by the sun), or by moving objects that have little infrared signature, such as a leaf being blown in the wind. Therefore, the snake combines information in the brain from two senses, to encode the type of object and where it is, in order to locate and respond to it appropriately.

While the infrared sense of these snakes is important in prey capture, some scientists have made the case that the pit organs effectively combine with the eyes to create an enhanced visual world. In practice, the infrared sense is not really part of the visual system, but does enable a 'picture' of the world and what is happening around it in space to be constructed by the snake. Moreover, different groups of only distantly related infra-red snakes have undergone parallel convergent evolution at the molecular level for a particular protein that is key to making the infrared receptors work (TRPA1). The gene that expresses this protein has been strongly modified by natural selection only in pit-bearing snakes, and not in species that lack an infrared sense.[12] The key protein in both types of snakes is sensitive to heat, including in temperature and pain sensations, but in pit-bearing snakes it has been modified to respond to lower temperatures and to better detect changes in the wider ambient environment (not dissimilar to the changes that have occurred in vampire bats). Again, we see how evolution often modifies existing genes for new tasks, and that the same solution may evolve independently in more than one group of animal. This is also the case with the electric sense of fish and echolocation in bats and dolphins, which we will return to later.

Sometimes animals combine senses in order to make the correct or most appropriate choice when they might behave in different ways entirely. Among many butterflies, individuals visit flowers to feed on nectar, often guided by the bright colours of the petals. Likewise, females seek out the green leaves of host plants on which to lay eggs. Studies of swallowtail butterflies (Papilio xuthus) have tested how likely individuals that have not seen flowers before are to approach coloured disks ('flowers') of red, green, blue, and yellow.[13] With just colour, for some reason most individuals show a preference for blue. But when presented alongside flower scents, females show an increased preference for red. In contrast, when disks are shown alongside the odour of host plants on which females lay their eggs, they then prefer green. The choice of colour is altered by the presence of different odour cues, and these changes make sense when we consider what features are used to find food or host plants.

Animal senses are spectacularly diverse, and we humans simply cannot know what it is like to perceive much of the information from the world to which other species respond. Instead, we have to rely on special technology and behavioural experiments to understand how they react to different sources of information. Even when we share the same sense as another species, it is often tuned very differently. For example, many animals see ultraviolet light, but we do not. The world we perceive is a product of our evolved sensory systems and brain, and this can be very different from many other organisms, which have evolved their own perceptions, dependent on their environments and needs. In the coming chapters, we will explore some wonderful examples of animal super senses, beginning with hearing.

2

SINGING RATS AND SONAR BATS

The greater wax moth (*Galleria mellonella*) is easily overlooked. Just a couple of centimetres long and dull in appearance, it is hardly a creature most of us would get immediately excited about. But looks can be deceptive, because this inconspicuous little animal has a number of remarkable characteristics. For one thing, it is a significant pest of honeybees, and is found over much of the world where honeybees exist. The female greater wax moth lays her eggs in the bees' nest, and the one or two larvae that successfully hatch out feed on the honey, wax, and pollen, causing damage to the hive in the process. The moths are probably able to invade and the larvae to live in the honeybee hives by mimicking the colony's smell and by sounding like bees, just as many other animals do. Wax moth larvae have recently drawn particular attention because they have been shown not only to eat types of plastic that are commonly used in packaging and plastic bags (and notoriously resistant to decomposition), but they may even be able to directly break the plastic down. This potential ability (still needing full verification) seems to stem from their normal practice of munching through wax, which has some properties similar to plastic.

For our purposes here, the most amazing thing about the greater wax moth is its hearing. It has been known for the past 20 years that the moths can hear very high-frequency, ultrasonic sounds, both the echolocation calls of bats and those made by the moths themselves. Wax moths beat their wings during mating, and this produces courtship sounds and helps to spread mating pheromones. When they hear the echolocation calls of bats, a major predator, the moths take evasive action. This is not unusual, and many moths and other insects do the same. But what is remarkable in the case of the wax moth is just how extreme its hearing is. Recent work has found that the moths can hear frequencies close to 300 kHz.[1] To put that in context, human hearing falls away at sounds beyond about 15–20 kHz, and even the most high-frequency echolocation calls of bats have not been found much above 210 kHz. Quite why such extreme hearing is needed is a mystery; it may be that some bat echolocation calls reach such extremes too and we have yet to measure them, and so the moths need to detect these. Or perhaps the moths communicate with extreme sounds to avoid attracting the unwanted attention of their enemies. Alternatively, it may simply be a by-product of the mechanics of how the moth's ears work. Regardless, this little moth illustrates just how limited our own hearing is. There is a whole acoustic landscape out there that we are missing.

The ability to hear depends on detecting sound waves of different intensity and frequency. Sound is a mechanical disturbance of a medium such as water or air. Sound waves travel from a source, perhaps a bird calling, and propagate through the medium (e.g. air) causing displacement of molecules and changes in pressure as they go (Figure 5). As different parts of a

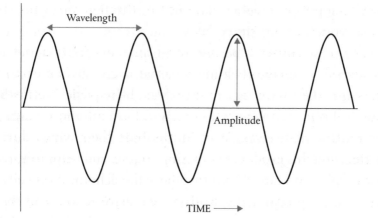

Figure 5. Sound travels through a medium as waves over time. A given sound wave has peaks and troughs, with the wavelength of a sound wave the distance between two peaks. Wavelength and frequency of sound, the latter being the number of waves per time period, are inversely related, such that sounds of higher frequencies (perceived as higher pitch) have larger wavelengths. The intensity of a sound, or its amplitude, relates to how loudly it is perceived.

sound wave pass, molecules either bunch together or spread out, increasing or decreasing the pressure, respectively. Sound waves also vary in intensity (how much energy they have and how loud they seem) as well as in frequency. Animals have evolved a wide range of hearing organs to detect features of sound, some responding more to changes in intensity, others more to changes in pressure. Ears have evolved numerous times independently, and they can occur in a range of structures and body locations. In insects, for example, ears can occur on the head, wing bases, legs, and a variety of other body parts. But wherever they are located, ultimately, hearing works by sound waves causing a mechanical disturbance of structures in a hearing organ, with different cells responding to particular frequency ranges.

Mammals like ourselves detect sound based on special cells and structures inside our ears. Sound passes from the external ear, though the middle ear and ear drum, to the inner ear. Here, sound waves arrive at the cochlea, which is covered with a special layer of cells, called the basilar membrane. We can imagine this membrane rolled out a bit like a carpet, with one end closer to where the sound waves enter the cochlea and the other end further away. Located on this surface are hair cells that are mechanically distorted by the vibrations when a sound wave passes, causing an electrical impulse in the sensory nerve cells connected to them. Different hair cells located along the length of the basilar membrane are responsive to particular frequencies, and these are grouped together in space. Cells closer to the start of the membrane are more sensitive to higher frequencies, whereas those further along its length are more sensitive to lower frequencies (low frequencies tend to penetrate farther). Conceptually, this is a kind of auditory map, in which the region of the basilar membrane that is most stimulated reveals which frequencies of sound were present. Other vertebrates have somewhat different structures in their ear, but all are quite similar in nature and in the mechanisms of sound reception. Changing the types of hair cells and their sensitivities to different sound frequencies alters the capacity of the auditory system to respond to different sources of sound information.

Hearing has numerous functions. For many animals, it is vital for detecting threats, such as an approaching predator. Just watch a deer for a few moments and you're sure to see it pause momentarily and move its external ears around, listening for potential danger. Hearing is also critical to a variety of other activities, from communicating territory ownership (in many birds, for instance) and trying to attract a mate (as with the wax moth above), to detecting prey in the undergrowth, and even sometimes in navigation. And in some animals, it can be highly specialized.

Few animals illustrate how remarkable the sense of hearing can be, or have taught us so much about how it works, as the barn owl (*Tyto alba*). They are adept at using their hearing to locate a sound source, for instance the noises coming from a scurrying mouse. Barn owls are a very widely distributed species; in fact, they occur all round the world in almost every region except Antarctica and some desert areas. In most locations, individuals hunt at night, though for some reason in the UK they are often diurnal. They search for small mammals like rodents and invertebrate prey moving along the ground. With their silent flight and ghost-like appearance, a barn owl hunting is a wonderful sight. The prey they search for produce subtle sounds as they rustle along the ground, which the owls use to locate and capture them. Being almost silent in flight is no accident. The owls have specially modified feathers that enable them to approach prey often undetected.

Back in the mid-twentieth century, Roger Payne started studying barn owls, first for his PhD at Cornell University, and then continuing well into the 1970s.[2] Payne demonstrated that owls can detect prey in total darkness based on sound cues alone. He conducted experiments in which he let mice run around in a room and allowed barn owls to hunt them. When the floor was covered in dry leaves the owls could easily capture the mice in complete darkness. However, when the floor was lined with sand they didn't do so well. Next, when Payne tied pieces of paper to the mice's tails, the owls captured them again (well, they attacked the paper in fact). Blocking the owl ears with ear plugs prevented them from locating the mice accurately.

Subsequent experiments added more detail. The use of speakers playing back sounds showed that the owls were particularly sensitive to sounds made by prey in the frequency range 4–9 kHz. By placing owls on a perch surrounded by a magnetic coil that was used to track head movements it was possible to measure the direction in which an owl was looking in both a vertical and a horizontal plane. At the same time, sound cues could be manipulated by, for instance, putting earplugs in one or both ears, or using headphones to play sound directly into the owl's ears at different times. All this work clearly showed that sound information was central to the owl's ability to find prey.

Payne's work, in addition to other classic research by Eric Knudsen and Masakazu Konishi,[3] established many of the key features that make barn owl hearing so effective. First, the face of the barn owl has a number of adaptations to facilitate sound reception. It is shaped like a dish—perfect for receiving sound and channelling it to the external ears. The dish is covered with special feathers that are acoustically 'transparent', and so, unlike normal feathers, they allow sound to reach the body beneath. In addition, the range of frequencies in which barn owl hearing was found to be most sensitive, 4–9 kHz, corresponds closely to the frequency range of sounds made by rustling prey on the ground. Another key anatomical feature is that the owls have ear openings that are asymmetrical. The left ear is higher up on the head, with the opening pointing downwards towards the ground. In contrast, the right ear is lower down and points more in a forward direction. The left ear is more sensitive to sound from below, and the right ear to noises in front of the owl.

So how does an owl locate its prey? It all comes down to how it processes information on the timing and intensity of sounds arriving at each ear. Early research showed that if you plug one of the owl's ears, the owl makes errors in locating prey in a vertical plane. Ear plugs change the intensity, but not the timing of sound. Because the left ear points towards the ground and the right ear straight ahead, the owl would normally use differences in the intensity of sound arriving at the two ears to know whether a prey item is below or in front of it. A running mouse directly below would present more intense sound cues to the left ear than the right, leading to greater stimulation of the hair cells in the left ear.

To calculate whether a sound is coming from the left or right, the owl uses differences in the timing of the arrival of sound to each ear. Imagine you are looking straight ahead and a sound originates straight in front of you. The sound will arrive at both ears at the same time. Now imagine that the sound occurs to your right; it will reach your right ear marginally earlier than your left ear. The owl can use such slight differences in timing to work out the location of a sound source in a horizontal plane.

The owl's auditory system combines all this information, enabling the owl to attack in the right location. Intensity and timing differences between the two ears are processed by the owl's brain in separate dedicated pathways in the auditory system. This is an example of 'parallel processing', and is commonly found in the processing of sensory information. Different pathways are used by sensory systems and brains to encode specific features of sensory information separately, before bringing them together at a later stage in another part of the brain. We will come back to how the owl does this in a moment, as well as why.

Calculating the differences in sound between one ear and another is not trivial. The ears are close together, and sound waves can travel at nearly 350 m a second, so sounds will arrive at the two ears just microseconds apart. Processing such sound information requires clever solutions to encoding fine differences at high accuracy, in order to enable the owl to capture small, moving prey. Recall that sound travels as a wave, which can be represented in terms of a repeating pattern of peaks and troughs. The phase of a wave is a measure of a specific point in the cycle of peak to trough. So, a particular phase might correspond to the lowest point in a trough or the position of the maximum peak, or anywhere in between. The remarkable thing about owls is that they have 'phase-locked' hair cells, which only respond to particular phases of the sound wave. That is, they fire at the peak or trough of a wave. This allows staggering accuracy in discriminating between sounds because the owl can then compare the tiny differences in sound waves arriving at the two ears in terms of their phase.

There is more to detecting time differences between the ears than phase, however. In principle, the owl's auditory system could encode differences in timing according to how many times the sensory nerve cells fire over a given period. But a nerve impulse lasts for milliseconds—an order of magnitude greater than sound differences between the ears, which occur over microseconds. So this approach can't work; the generation of nerve impulses is too slow and they can't be produced quickly enough. But the owl has evolved a neat trick, one that engineers have come up with independently—a 'coincidence detector'. This works on the basis that the speed at which nerve impulses travel along a nerve fibre is constant (assuming the fibres are the same width, plus certain other factors we need not worry about here). All things being equal, it takes a nerve impulse a longer time to traverse a longer nerve fibre. The owl auditory system uses this fact to its advantage by having special nerve cells, the coincidence detectors, that produce a response only when they receive simultaneous stimulation from nerve fibres from the left and right side of the body. In the owl's auditory system, there is an array of these cells, some receiving impulses from nerve fibres that are longer from the left than the right ear, and some the other way round. This means that such arrays of cells will only be activated based on specific timing differences between the left and right ears—whichever coincidence detector fires corresponds to a specific time difference between a sound arriving at the left and at the right ear. It is a very neat solution that enables extremely small and precise sound differences to be encoded, even allowing the owl to detect timing differences in the order of microseconds.[4]

When it comes to detecting differences in intensity, things are rather more straightforward. Sounds will be more intense when the source is closer and when it is more directly facing one ear, and this in turn will produce a greater stimulation in the auditory system. Given that the ear openings on the left and right side point in different directions, they will be more sensitive to the intensity of sounds arriving in different vertical planes. So, depending on which ear is more strongly activated, and the difference between the ears, this tells the owl about the direction to the sound source in front and below it.

The barn owl's hearing becomes even more impressive when we consider how it utilizes such sound information to bring about a behavioural response. The two pathways processing intensity and timing differences converge in a part of the brain called the inferior colliculus. This is where the owl combines information about an object's position in space. To do so, its brain uses another sophisticated processing trick, one that also occurs widely in sensory systems: the use of receptive fields to create a spatial map. Across the inferior colliculus are cells— receptive fields—that are only activated when they receive specific stimulation from the timing and intensity pathways that correspond to a particular point in space, such as directly ahead of the owl or downwards on the left-hand side. This is an auditory map, in that the spatial position of any source of a sound in the external world has corresponding cells with a receptive field in the owl's brain. When a specific cell fires, the owl can rapidly orientate to the location of that sound source, and target the source quickly and accurately when hunting.

Barn owls have a truly wonderful ability to use their hearing to find food in the darkness. In fact, as we know from Chapter 1, they also use their vision and have some clever tricks in their brain to combine hearing and vision to become even more adept at pinpointing targets. And they have a further ability as well, something that humans could really benefit from. When researchers tested the hearing of young (two-year-old) and older (thirteen-year-old) owls to sounds of different frequencies and intensities they found little difference between the two groups. The same was true when they tested the hearing of one individual several times over twenty-three years.[5] Barn owls just don't seem to lose their hearing with age like we do. It is known that some birds can regenerate the hair cells in their ears that are used to detect sounds, and presumably it is this ability that protects owls and other birds from suffering deteriorating hearing as they get older.

Our hearing range (though age-dependent!) covers roughly 15 to 20,000 Hz. However, many mammals (especially over 1,000 species of bat), insects, and other groups detect ultrasonic frequencies well above this range (we have already encountered the wax moth). The fact that many animals hear ultrasonic sounds means that we have often tended to miss important aspects of their behaviour and communication, at least until we use specialist equipment and experiments to uncover it. Fascinating aspects of the biology of many species have been discovered when we do this work, not least in a variety of rodents. Mice and rats are well known to use smell in communication. In fact, much scientific investigation has shown how mice can discriminate between relatives based on their smell. Many rodents also use odours in things like marking territories and providing information regarding reproductive state. Beyond chemicals, it is now known that numerous species communicate with sound and can hear ultrasonic frequencies, and they make use of them in a variety of contexts.

Research from the 1950s and '60s had begun to document ultrasonic calls made by animals other than bats, especially baby mice and rats when under stress (when they were cold or hungry).[6] Then, a series of separate studies in the mid to late '60s by Eliane Noirot at Cambridge University and Gillian Sales at King's College London began to shed light on the widespread nature and functions of this phenomenon. Noirot studied young albino mice and found that pups made calls within a particular time window of their development. The behaviour begins after around four days, when the ears open, and stops at around 13 days, when the eyes open. The calls tend to occur when mice pups are retrieved back to the nest by the mother, or as distress calls when pups are left in isolation. Noirot documented very similar results in albino rat pups, where the calls act to guide parents towards lost pups and help them to retrieve the pups and return them to the nest. Baby rodents appear to use the calls to induce care from the mother. Experiments directly playing back the calls of young confirmed that mothers do indeed respond to pup vocalizations, in both mice and rats.

We now know that rodents produce ultrasonic calls, called ultrasonic vocalizations (USVs) even into adulthood, and USVs occur widely across a range of species, with much variation in the use and structure of the calls depending on the species, lab strain, and age. It seems that rats, but not mice, make USVs in aggressive encounters, while both make USVs during mating behaviour. The use of ultrasonic calls in many species and contexts is now widely documented, ranging beyond mice and rats to creatures such as lemmings, gerbils, hamsters, and voles.

Different types of vocalizations also seem to be used for particular tasks. In rats (*Rattus norvegicus*), short USV calls (of about 50 kHz) have been linked to aggressive encounters, whereas long USVs (25 kHz) are used in submission, acting to reduce aggression. These distinct call types appear to mediate dominance and social interactions. Various species also use USVs widely during mating, produced primarily by males to entice females.

By the end of the 1970s, our understanding of how rodents use sound to communicate had completely changed. After this initial spate of work, the study of rodent USVs ticked along until the past decade or so, when a new research drive on USVs shed light on just how sophisticated rodent vocal communication is. It turns out that the calls aren't simple sounds of one pitch that are 'turned on or off' but that they are intricately structured. Studies of the USVs of male mice have revealed that the vocalizations consist of elaborate patterns of repeating syllables and phrases (sequences of syllables in close succession), amazingly similar to the patterns in birdsong (Figure 6).[7] The rats seem to be singing to females as part of their mating behaviour. What's more, even in highly genetically similar males, the calls are different, suggesting that the rats' songs convey some sort of information, perhaps about the quality of the male. This discovery has drawn much interest into rodent USVs, including those researchers studying song and vocal communication beyond humans and birds.

A broad picture of rat and mouse USVs is now emerging. Rats generally produce three kinds of calls: a call at roughly 40 kHz by pups when separated from their mothers; a long 22–25 kHz call by adults when in distress or pain, when under the threat of a predator, or when defeated in an aggressive encounter; and, finally, a short 50 kHz call when in a 'positive state', including during play, sexual encounters, and when doing well in aggressive bouts—that is, in situations when there may be some sort of

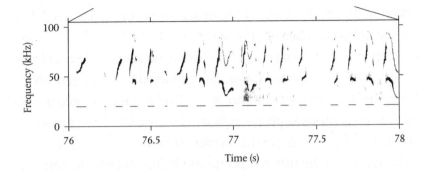

Figure 6. Ultrasonic songs (above 20 kHz in frequency) of male mice produced when they detect the chemical cues of a female. Songs have distinct structures and syllables that repeat frequently over time, characteristic of song in other animals such as birds.

reward.[8] This helps to explain why rats handled by familiar humans (e.g. pet owners) emit 50 kHz calls, whereas those handled by unfamiliar humans emit 22 kHz calls. When rats are given 'experimenter-administered manual stimulation', that is, when they are tickled by a human, they seem to enter a positive state, emitting 50 kHz calls which we might consider analogous to laughter.[9] Adult mice use complex songs with syllables and phrases during social interactions, especially mating encounters but also in territorial behaviour. Both males and females emit USVs, but while male mice call mostly towards females, females tend to use USVs in searching for pups or in interacting with other females. Mice also appear to emit USV calls when exploring novel environments.

It isn't enough to uncover the presence of songs in rats and mice during mating; biologists need evidence that they play a role in mating behaviour to confirm they have a function. Sure enough, females are drawn to the songs of male mice played back to them, and male mice start to sing in response to female mice or if their smell is present. Female mice also show a preference for the USVs of unfamiliar males, which is probably because these males are less likely to be closely related to them. It seems to be a way of avoiding inbreeding.

Male mice can also modify the nature of their song characteristics. When they sense the presence of females in the vicinity, by smelling female urine, males sing louder and more complex songs, with more sophisticated use of syllables and sequences. Yet when they are in close proximity to a female, they use longer and simpler syllables. If female mice are played the songs of different males, they prefer complex songs, just as many female birds do. It seems that males use complex songs to attract females in the first place, and then simpler versions during actual courtship and mating. The picture of rodent ultrasonic songs is developing quickly. Male calls also seem linked to certain hormones and behaviours associated with dominance, which suggests that aspects of male song may encode fitness. In turn, females respond to the presence of specific proteins found in male urine with increased vocalizations too, helping to guide reciprocal communication between the sexes.

There is one more fascinating feature of vocal communication in rodents that has just come to light, which has similarities with how we talk to one another. When we converse, we tend to speak in turns. We all know how rude it seems when someone talks over us, and apart from anything else, this makes it harder to interact. If we say our piece and then listen to the response, we can focus our attention much better. Some rodents, it seems, do the same thing. The individuals of a small species of mouse that lives in the cloud forests of Central America, the appropriately named Alston's singing mouse (*Scotinomys teguina*), take turns in vocalizing to one another.[10] In the lab, the scientists working on the mice had observed that when their cages were close together the mice seemed to engage in conversation. Both males and females produce distinctive songs covering a range of frequencies, comprising sets of notes. The process of taking turns to sing, and of controlling the speed of the interaction, is managed by a specific area of the brain, the motor cortex, even though this part of the brain is not apparently needed for producing song in the first place. Such behaviour, of taking turns to communicate, is common in other animals too, and indeed we would expect that to be the case, though interestingly it had not been found in conventional laboratory mice.

Most studies of rat and mice communication have involved laboratory strains but it is certainly also the case that wild-type and free-living rodents produce USVs too, although the significance of these in wild animals and under natural conditions is yet to be properly explored. It is quite likely, given the levels of inbreeding in lab stocks of rodents (which tends to result in the loss of natural variation of traits—like eye size in flies) that wild animals are even more diverse in their USVs. Unlike birds, among which song is partly learnt through exposure to a tutor, mouse song appears to be more strongly genetically controlled.

Quite why rodents use ultrasounds rather than lower frequencies isn't really known. Sales suggested that it may be merely a function of their small size, and that it is unlikely to be a means of avoiding detection by predators since many of these, like cats, can hear ultrasounds too. Ultrasound may not be 'special' in any way, other than that we humans can't hear it. We can only wonder at the songs and chatter happening among so many animals around us, of which we are quite unaware.

Of all animals, the group that must surely have the most remarkable and sophisticated hearing is the bats. Many bat species can echolocate using sounds that they produce themselves. They are among a few select groups of animals which don't just passively detect information from the environment, but actively emit signals and obtain information by measuring changes in the returning component. Not all species and type of bat do this, and the ability is more rudimentary in some bats, but it occurs in over 1,000 species. In those that possess it, it is often highly sophisticated. Echolocation is an example of an active sense, in which an animal generates signals that interact with the environment and is able to deduce information about the world around it from the returning response.

Early discoveries about bat echolocation were rather extreme in their design for the unfortunate bats. The Italian Lazzaro Spallanzani first noted in 1794 that blinded bats could still negotiate obstacle courses, whereas the Swiss naturalist Charles Jurine found that bats with their ears blocked with wax would crash into wires. But many doubted the existence of bat echolocation until more definitive proof came in the late 1930s, from the Harvard scientist Donald Griffin, working with the physicist George Washington Pierce, who had developed a device for converting ultrasounds into human audible frequencies. They showed not only that bats produce ultrasonic calls, but that they modified these while attacking prey.

In bats, echolocation is primarily used for detecting prey and for navigating through an environment. It involves high-frequency echolocation calls, sometimes up to and over 200 kHz and at very high intensity, even beyond 140 decibels. As a comparison, shouting loudly at someone from about 1 metre away produces around 80 decibels. It's a good job we don't hear ultrasonic calls or else walking out at night would be a deafening experience. Actually, this is a problem for the bats too. Think of a bat flying in a wide open space, needing to detect and locate a small flying moth; the proportion of sound reflected back off the moth to the bat is going to be tiny, so blasting out lots of sound can be essential. To avoid deafening themselves, many bats shut down their middle ear when emitting calls. In fact, bats have evolved a whole suite of anatomical and neurological adaptations to produce and receive sound. The faces of many species are a remarkable thing to behold in themselves. They vary greatly, typically having enormous external ears and elaborate mouth and nose appendages to emit sound.

The calls produced by bats are highly refined and specialized for particular tasks and aspects of each species' ecology. But we can broadly group bat calls into two main categories—constant frequency (CF), and frequency modulated (FM)—and each type has specific advantages. CF calls, as the name suggests, are relatively long (10–100 ms) and mostly within one frequency band, and are emitted with very high intensity. They tend to be used for detecting objects in wide open spaces, because the high intensity of the call allows a good chance of an echo returning with sufficient power to be detected. CF calls are good for finding a small moth flying in an open area like a field. FM calls are short (1–10 ms) sweeps of many frequencies, and tend to give many different pieces of information corresponding to each frequency, so they are good for deriving accurate and refined information and in contexts such as flying in cluttered environments (Figure 7). In reality, the distinction between the two is fuzzy. Most CF calls actually begin or end with an FM component.

To understand the sophistication of bat hearing and how echolocation is used, we can think about how a bat would capture a prey item. Imagine a bat in flight through an environment, hunting a small moth that is also flying in the air. The first thing the bat has to do is detect where in space the moth is; that is, its location in a horizontal and vertical plane. To do this, it uses much the same approach as the barn owl, only with the bat itself emitting the sound and listening to the returning echoes. To work out the position of the moth in the horizontal plane, the

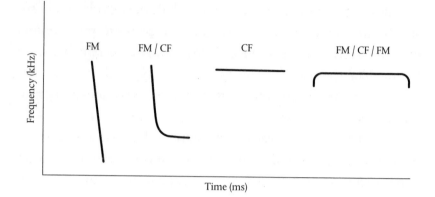

Figure 7. The echolocation calls of bats can be broadly categorized into a few types. Frequency modulated (FM) calls involve downward sweeps of frequency as the call is produced and tend to be relatively short in duration. They tend to be useful for gathering lots of information at different samples (frequencies) in cluttered environments. Constant frequency (CF) calls tend to be high energy calls occurring at about the same frequency over a longer time period, and are especially valuable in detecting objects in wide open spaces and for information such as velocity. In practice, most CF calls tend to be accompanied by a FM component (CF-FM).

bat calculates differences in timing of the echoes arriving between the right and left ears. For elevation (position in the vertical plane), the bat listens to differences in sound intensity above and below its head, and it can often do this more effectively by having the ability to move its external ears (pinnae) to point to and receive sounds from different directions.

Now, because the bat and moth are in motion, the task of capturing the moth is much harder. The bat needs to know other information, such as how far away they are from one another and how fast they are each flying. To calculate distance (or the range to the target) the bat can use the time delay between when it emits its call and when the echo returns—a longer delay means that the object is further away since the sound waves take longer to travel out and then back from the moth. Here, bats often use FM call information, because these call types give lots of information at different frequencies, and so provide highly accurate distance data.

To calculate speed, many bat species use a phenomenon known as Doppler shift. This is a change in the frequency of sound reaching a receiver resulting from the movement of the receiver itself, the sound source, or both. We encounter this effect in the rising pitch of the siren of an approaching ambulance or police car, while one that is moving away from us drops in pitch. We experience the shift in frequency because when the source of the sound is approaching us, the sound waves reach us at a greater rate because of the vehicle's own motion, which is equivalent to a higher frequency. By contrast, when the sound source is heading away from us the sound waves arrive at a slowing rate again because of the vehicle's motion, so we hear a lowering of pitch. When bats approach an object very quickly, they perceive a rise in the frequencies of the returning echoes. Conversely, an object moving away from them will cause negative Doppler shifts, or lower sound frequencies. To detect such information, CF calls are ideal because they are high energy and occur at one frequency band, so the bat can relatively easily detect the change in frequency between the call it sent out and the returning echo, and hence the Doppler shift, and from that calculate velocity.

This is all well and good, but before a bat attacks it would want to know what it is that it is actually hunting. Here, things get really clever. For a start, the size of the object can reveal much about what it is—a big moth or a tiny mosquito, for example. All things being equal, big objects reflect more sound back, so a larger moth would reflect echoes of greater sound intensity. But it's not that simple, because an object a long way away would reflect much less sound than one close by. In other words, the intensity of the returning echo could be the same for a small object close to the bat, or a big object far away. So, one thing bats do is to combine the intensity of the returning echo with the echo delay, which reveals distance, in order to work out object size. Some bats have another trick: they use the width of the sound beam that returns; but we won't go into that here.

Perhaps the most remarkable thing bats can do with sound is to identify fine-grained information about a target. They can work out the texture of an object based on minute features of the returning echoes (FM calls are especially good for this). They can tell the difference between the smooth and shiny wing cases and bodies of a beetle, compared to the textured scales of a moth's wings. Bats can even work out how fast an insect is beating its wings. Consider a bat approaching an insect side on. When the insect raises its wings from below, they will first move upwards and slightly towards the bat, before being held horizontally. Then the wings will continue upwards and move away from the bat. When the wing beat goes down, the opposite will be true; the motion of the wings causes them to move slightly towards and away from the bat in a characteristic speed pattern. This produces tiny changes in sound frequency via Doppler shift, which manifest as minute oscillations in the returning echoes. Incredibly, bats can use CF calls to work out the Doppler shift changes caused by wing beat frequency, and use this to determine what they are hunting.[11] Some insects such as mosquitoes have very rapid wing beats, whereas a lumbering big moth would be much slower, and so the frequency of oscillation of the echo reveals the type of insect.

Bats, then, can use their echolocation to calculate distance, velocity, 3D location, and direction to a prey item, as well as its size, texture, and wing beat properties. If that isn't a remarkable sensory ability, then nothing is. Somehow, bats must process all this information in their auditory system and brain. Fortunately, there has been a great deal of research into how they do this.

To understand how the bat auditory system processes echo information, we again come back to the different uses of FM and CF calls. In a region of the bat's brain called the auditory cortex, specific areas are dedicated to processing information from these two types of call. An FM-FM area encodes differences in the timing between the emitted call and the returning echo, which relates to the distance the bat is from the object. Bats have the same problem here as barn owls, in that encoding differences in timing that can be very small can't be done based on the frequencies at which nerve cells can generate impulses—nerve cells are too slow. Sure enough, as in the barn owl, work on the little brown bat (*Myotis lucifugus*), a small mouse-eared bat that lives in North America and hunts insects, has shown that bats too possess coincidence detectors. Its coincidence detector cells detect variation in time differences between calls and echoes.[12] These arrays of cells, called delay-sensitive neurons, are only activated when specific time differences arise between a call and echo. When the bat calls, it triggers an internal signal along a neural pathway. The point in time at which the echo returns is detected, and then 'meets up' with the internal impulse, which corresponds to a specific delay-sensitive neuron. This in turn encodes a particular time difference, and, therefore, real distance. These delay-sensitive nerve cells are found in many other bats (they have been well studied in horseshoe bats too, for example). The cells form an array in the FM-FM area that produce a region of different echo delays (Figure 8). The amazing thing is that this is another spatial map of sorts, because a specific part of the FM-FM area corresponds to specific echo delays in actual millisecond times, and therefore real distances in space.

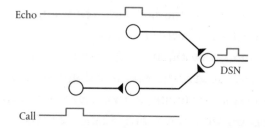

Figure 8. Echolocating bats can encode the differences in timing between an emitted echo and a returning sound through delay-sensitive neurons in the brain. They use this to measure the distance to a target. When a call is made, a signal begins to move along an internal pathway, and a second pathway is triggered when the returning echo is detected. The two impulses meet at a particular delay-sensitive neuron (DSN), which relates to a specific timing difference.

To detect the subtle changes in frequency of CF calls due to Doppler shift, associated with the speed at which the prey is approaching or moving away, many bats have evolved greatly enhanced auditory systems that are highly sensitive to fine differences within a narrow frequency range. Mustache bats (*Pteronotus parnellii*) are a widespread insectivorous species found across the Americas. They are relatively large bats, with a wing length of 5–6 cm, and a mouth shape in which the lips are arranged into a funnel. In this species, well studied for its echolocation, there is a strong focus on sound reception around 61–62 kHz, which is very close to the frequency of its calls and likely echoes. In fact, 50 per cent of the hair cells in its cochlea are sensitive to sounds between 50–74 kHz.[13] And there is another specialist area in the auditory cortex, the 'Doppler shift constant frequency' or DSCF area, which specifically encodes changes in frequency. Here, 50 per cent of all the nerve cells encode frequencies of just 61–63 kHz, and they are again arranged spatially in terms of frequency. This extreme specialization allows the bat to detect tiny changes in frequency between the call and echo, and the oscillations caused by insect wing beats.

We now also know much more about the underlying evolution and genes enabling bats' echolocation powers. In mammals, sound production (including human speech) is controlled by a gene called *FoxP2*, and this is also a key gene in the USVs of mice. In most mammal groups there is little variation in the gene (it is said to be highly conserved), but in bats it has been modified extensively by evolution and is highly variable. This has allowed bats to greatly change the specific call types and frequencies they use as they spread into new habitats and evolved new ways of living. The ability to hear, and specifically the sensitivity of cochlea hair cells to different frequencies, is underpinned by another gene, called *Prestin*. In bats, mutations to this gene, especially in those species that use predominantly CF calls, have enabled them to fine-tune their auditory sensitivity to their environment and way of life.[14]

I do not want to give the impression that bats use their echolocation solely for hunting insects. Far from it. Although this is common, bats are extremely diverse and many species use echolocation to find a great variety of other sources of food as well as shelter. Some bats are known to hunt frogs by eavesdropping on the male frog's mating calls. When the frogs realize that they are under attack they often stop calling, in the hope that they won't be located. But when sitting in a pond the water ripples from any movement continue to propagate across the water surface even once the frogs have gone silent. Bats like the fringe-lipped bat (*Trachops cirrhosus*), a species from Central and South America that is expert at eating amphibian prey, use their echolocation to detect the ripples and their original source. Often, these bats listen in to the calls made by frogs to find them (rather like the parasitoid flies), but if the frogs go silent they switch to echolocation mode and search for characteristic water ripples in ponds (Figure 9). Other bats visit flowers for nectar rewards, acting as pollinators in the process, and use their echolocation calls to find special structures that have evolved on the plants to enable the bat to locate them easily in the forest.

Bats are the second most widespread group of mammal (after rodents) and have been around for millions of years. It's not surprising therefore that they have placed selection pressure on the

Figure 9. Some bats specialize on eating amphibians and can locate their prey by listening in (eavesdropping) on the calls that the frogs make. Once frogs sense a risk of attack they may stop calling, but bats such as the fringe-lipped bat (*Trachops cirrhosus*) can still locate them by detecting the ripples of water that continue to travel across the pond using their echolocation. Here, a fringe-lipped bat has captured a Fitzinger's robber frog (*Eleutherodactylus fitzingeri*) in Panama.

organisms they hunt to evolve defences, in a continuing arms race. Many of the hearing organs that exist in insects have either evolved, or have at least been modified, under selection pressure to detect the echolocation calls of bats, so that the insect can take evasive action. We already encountered this with the wax moth and its remarkable sensitivity to ultrasonic sound. In some moths, the ability to hear ultrasound has been extended further, with males now producing ultrasonic mating calls which the females can readily detect. So not only have bats led to a range of anti-predator defences in insects, they have also played a part in driving the evolution of hearing and song, and new communication strategies in these completely different animals.

It's not just bats that emit sound and use it as a form of sonar. Over seventy species of toothed whales, particularly dolphins and porpoises, do much the same. They use their sonar abilities to communicate with one another in social groups, and to coordinate and hunt for food such as fish. In their case, sound is produced in the nose and transferred into the water by means of an 'acoustic melon' on the forehead which directs it into a more focused forward beam. Returning echoes are then picked up by the lower jaw and pass to the inner ear. Like bats, dolphins use echolocation for detecting objects too, but for them it tends to have greater importance in communication.

Bats and dolphins illustrate some remarkable examples of convergent evolution in the use and production of sound, and we now know much about how they do this. Relatively recent research has shown that there is not only convergence in the use of echolocation in these groups, but also convergent evolution at the genetic/molecular level underpinning the changes.[15] If you construct a family (phylogenetic) tree based on large areas of the genome of different groups of bats (both echolocating and non-echolocating), as well as of the echolocating dolphins and porpoises plus baleen whales (which have no echolocation), you arrive at the expected outcome: all bats cluster together into a true group, being closely related to one another, and the whales and dolphins cluster into a second, separate group. It would be really odd if some bats were more closely related to any whale than to other bats. The twist comes when you construct a family tree based on only the *Prestin* gene, which underlies sound frequency detection. Now, you produce a tree in which echolocating bats and echolocating dolphins cluster *together*, and separately from the other non-echolocating species (Figure 10). This tells us

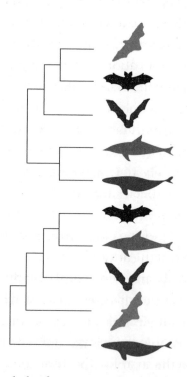

Figure 10. Bats and whales form two separate independent evolutionary groups (top). However, when a phylogenetic (family) tree is made based only on sequences of the *Prestin* gene (bottom) in these groups then echolocating bats cluster alongside echolocating dolphins, showing that the same changes to the gene have occurred in each group to underpin echolocation.

that the same, or similar, mutations have occurred in the *Prestin* gene in both dolphins and echolocating bats. Convergent evolution can therefore occur at the molecular level as well as in the outward appearance and behaviour of an animal. This was an important finding because conventional wisdom had been that convergent evolution would pretty much always have a different molecular/genetic basis across unrelated species. This will surely often be true as well, but echolocation shows us how similar changes can occur at the gene level in independent evolutionary events. Perhaps this is not surprising if evolution has limited genetic material to act on in the first place, and we will return to this in Chapter 4 when we discuss electric fish.

Bats and dolphins are by far the most advanced with regards to how animals use echolocation. However, the ability is also found in some other small mammals such as certain shrews, and even humans can learn to echolocate using parts of the brain that are normally used for vision, albeit in a very crude way. Most notably though, echolocation occurs in a relatively rudimentary form in some birds, specifically the South American oilbird (*Steatornis caripensis*) and some species of Indo-Pacific swiftlets. The birds use frequencies audible to humans, around 1–15 kHz, in the form of simple 'clicks'. Exploring caves in Venezuela in 1817, the great naturalist and polymath Alexander von Humboldt noted the use of clicking by oilbirds, and a great deal of early work on the echolocation of birds was performed by Griffin, who pioneered studies of bat ultrasonic echolocation. Despite its comparative simplicity, avian echolocation has evolved several times independently and occurs in at least sixteen species. I have encountered swiftlets coming in and out of caves in Borneo and it is certainly possible to hear the clicks they make for echolocation. They use this ability primarily for navigating through caves and in darkness, and potentially in communication. It's possible that oilbirds use clicks to locate fruit to eat as well, but swiftlets eat small insects, and their echolocation is too rudimentary for use in targeting them. Nonetheless, the hearing of echolocating birds seems broadly tuned to pick up effectively the echolocation frequencies that they produce.

The acoustic world is rich and diverse and used by animals in performing most if not all of the main activities of their lives: find food, avoid predators, find and attract mates, mediate social interactions, and navigate through the environment. Understanding this has taught us a great deal about how sensory systems and brains work. Sound is clearly of great importance to humans—after all, our primarily means of communicating and the basis of our language is in this modality—and we too often rely on it to detect threats and danger. We delight in the sounds of the natural world, perhaps birdsong most of all. Yet so much of the sound world of other animals is hidden from us, and we can only marvel at the staggering levels of sophistication that many species have in using sound themselves. In Chapter 3, we turn to vision, a sense with which we are also intimately familiar, but one that also showcases remarkable variety in nature.

CHAPTER

3

FOR MY EYES ONLY

On a starlit night, in South Africa, a beetle rolls a ball of dung away from the large pile it had been visiting, manoeuvring it back to a safe distance. Here it can escape the competition from other beetles and eat its dung in peace. The nocturnal dung beetle (*Scarabaeus satyrus*) is able to move away, with its precious load, for over a metre in a straight line, some fifty times its body length, over the uneven terrain. This matters because if its path curved round in a circle it might simply return to where it started, back with all the competition. The way in which the beetle is able to achieve this feat of travelling so far in a straight line is quite remarkable. In 2013, Marie Dacke from Lund University in Sweden and her colleagues showed that the beetles orientate at night by the stars.[1] And not just any stars, but specifically the light from the Milky Way. Under cloudy skies, the beetles cannot do this, and when the researchers let the beetles transport dung in a planetarium under different star patterns they could only do so when the Milky Way was present. A later study by the team used an ingenious technique of mimicking the light of the Milky Way in a planetarium with LED lights. This revealed that the beetles responded to subtle differences in brightness between parts of the Milky Way and the surrounding sky. Other species orientate using light from the sun or the moon, for transporting dung and moving around the environment; and, more broadly, many other groups of animals, such as ants, also use cues from the sky and visual landmarks in their habitat. In each case, they illustrate how crucial vision is to survival, and how individuals respond to a diversity of information in the world around them. In this chapter, we will explore how vision is used by animals, and the stunning diversity in ways of seeing.

In true vision, the light-receiving organ (the eye) should encode patterns of light and dark across the visual scene, rather than simply responding to the presence and overall intensity of light (phototaxis), which even some unicellular organisms can do. This may amount to just a blurry black and white image, but in more sophisticated examples it can produce images of high detail and a diversity of colours. Eyes have evolved many times independently in nature. As we would therefore expect, many eye types exist—over ten broad categories in fact, far too many to discuss them all here. A wide variety of animals, especially vertebrates like humans and birds, and some invertebrates such as cuttlefish and jumping spiders, have what can broadly be referred to as a 'camera-type' eye. This consists of a single opening and often a lens structure for focusing light onto an array of light-sensitive cells, or photoreceptors, at the back. Such eyes generally provide acute vision for seeing fine details and pattern. Another important type is the compound eye, of which there are several kinds. They are found in many invertebrates from flies to prawns. Instead of one lens, the compound eye is composed of thousands of tiny lenses or structures called ommatidia, within each of which lies a set of photoreceptors. The compound eye provides lots of little images that can be combined into a bigger picture, albeit a low-resolution image of the world.

The light that humans can see, what we call 'visible light', is part of the much larger electromagnetic spectrum, which spans extremely high-frequency (short wavelength) gamma rays at one end, and low-frequency (long wavelength) radio waves at the other. Somewhere in the middle is visible light, of which humans can see approximately 400 nm ('blue') through to around 750 nm ('red') light, though the precise range depends on a number of factors, not least light intensity. Many animals perceive light outside this window.

Before we come to our first animal group of this chapter, let's pause to consider just what colour and visible light really means to humans and other animals. Much of what we see in the world is simply made up of differences in brightness; this is why black and white TV was so successful before colour devices came along. However, there are problems with relying on brightness to read the world. For one thing, it can be unreliable, since it tends to vary over time as weather conditions change, and over space, for instance with patches of dappled light in forests or reflections in water. Brightness tends to be a noisy signal to use if you want to identify something. Colour on the other hand is often thought to be more stable. The colours of objects do not tend to fluctuate so much over time, weather conditions, or habitat. One of the main reasons for the existence of colour vision is believed to be the advantage it brings in detecting fruit and other food in the environment, and certainly most flowers rely heavily on colour to attract pollinators against the backdrop of green vegetation.

The range of light wavelengths that animals see depends on their visual system. Many species have the ability to see the colour of objects. But what is 'colour'? In fact, when we say things like 'see an object's colour', we are not strictly being accurate because objects in nature do not have colour. Isaac Newton, during his famous work on light and optics, split white light through a prism into different wavelengths, creating a rainbow of perceived colours.[2] Yet he knew that these wavebands of light did not have their own hues: 'the rays of light, so to speak are not coloured'. Instead, colour is a product of our visual system and brain, which gives us a sensation of colour when particular wavelengths of light reach our eyes. A lemon is not yellow and a cucumber is not green, except in our mind. Biologists too have long understood this. Alfred Russel Wallace, the great Victorian scientist who co-discovered natural selection and spent much time studying colour in nature, remarked: 'What we term colour is a subjective phenomenon, due to the constitution of our mind and nervous system; while objectively it consists of light-vibrations of different wavelengths emitted by, or reflected from, various objects.'[3]

Quite how the brain conjures up the rich sensations of colour (or indeed any other sensations) we experience, the so-called *qualia*, remains a deep puzzle associated with the problem of consciousness, which has taxed philosophers and neuroscientists for decades. But we do know quite a bit about the visual system. Seeing colours at all requires the ability to discriminate between different parts of the spectrum, and establishing the colour of a particular object involves comparing the amounts of different light wavelengths present. To do this, you need to have at least two types of photoreceptor that are sensitive to different wavebands of light, and, conventionally, connections in the visual system that compare the outputs between these.

It is easiest to begin by thinking about colour perception in humans, because that is what we are familiar with. A person with 'normal' colour vision has three types of cell called cones, each sensitive to a different range of wavelengths. The 'blue' cone detects shortwave light between about 400 and 500 nm, being most sensitive to light of 420 nm. The 'green' cone is sensitive to light between around 450 and 650 nm, especially 530 nm. Finally, the 'red' cone detects light from about 475 nm to beyond 700 nm, peaking in sensitivity at about 560 nm.

In the early 1800s, Thomas Young, a British physician and polymath, developed a theory to explain how humans see colour, now called the 'trichromatic' theory (tri, as in three cones or colours). The argument is that the colours we see depend on the relative stimulation of each cone type. If the blue cone is strongly stimulated but the others are not, then we see blue. Similarly if the red cone is strongly stimulated compared to the others, we see red. When both red and blue cones are stimulated, we see purple. And so on. The trichromatic theory explains why most of the colours a person can see can be matched by a combination of three monochromatic lights.

The trichromatic theory goes a long way to explain our colour perception, but there are some problems. First, humans have not three but four primary colours: red, green, blue, and yellow. Second, there are types of colours that we never describe, specifically blueish-yellow and reddish-green colours—these simply don't exist. Instead, the sensation of some colours inhibits the perception of other 'opposing' colours. Finally, we also sometimes see illusions of colour, called 'image after effects'. For example, if you stare at a blue object, then replace this with a sheet of white paper, you will sometimes see an afterimage of the object in yellow. None of these aspects of our vision can be explained by the trichromatic theory. The solution came in 1892, when Ewald Hering, a German physiologist, proposed his 'opponent colour' theory. Hering argued that colour perception stemmed from pairs of colours being grouped into antagonistic neural pathways, such that when we see one colour it inhibits the pathway we would use to see the opposing colour type. In humans, it so happens that we have such channels, one for seeing red or green, and one for seeing blue or yellow (with yellow being the combination of red and green light). This explains why we never see reddish-green colours because if we see red, the pathways prevent us from seeing green. It also explains why we see colour after effects. In the example above, looking at something blue inhibits our yellow pathway—remove the blue and the inhibition is gone and the yellow pathway 'bounces back' strongly, and temporarily we see a sensation of yellow.

Humans are not particularly representative of colour vision in nature, and that's partly because this ability varies greatly across animals. Honeybees make a good example, partly because, like humans, they are also trichromatic and have good colour vision, but not in the way we know it. Bees have long been a subject of attention for work on colour vision, and the presence of colour vision and its importance for foraging was shown in work by Charles Turner in 1910, a few years prior to studies by Karl von Frisch.[4] Bees have three types of cell in their eye used for colour discrimination. However, things deviate when we consider what visible light means to them. Compared to humans, the bee's visual system is shifted to shorter wavelengths. Unlike us, they do not see longer wavelengths of red light well, but their visual range extends into the ultraviolet (predominantly very short wavelengths between about 300 and 400 nm). Our blue cones are somewhat sensitive to UV light but our lens does not transmit it and so we don't detect it. And we also lack a dedicated receptor to pick up and analyse UV separately from blue. In the earlier days of eye surgery, when people had artificial lenses put in, these were often transparent to UV. The bearers sometimes reported a higher sensation of blue because their blue cones were now also receiving UV light. Bees have a dedicated UV receptor, and then one for seeing shortwave ('blue') and mediumwave ('green') light.

So their world is still likely rich in colour, but not those that we see. Other animals deviate more substantially, in that they have either more or fewer receptors used in colour vision, and hence different 'dimensions' of colour perception (Plate 1), but we will return to that later. For the moment, let's consider what differences in the perception of the various parts of the spectrum means to animals, in particular in the case of ultraviolet light and jumping spiders.

Whereas many spiders rely on a remarkable sense of touch to respond to threats, court mates, and locate prey, one particular group of spiders depends more on their vision. Jumping spiders are exceptional hunters—they possess a keen sense of sight, used to detect and then creep up and pounce on their prey. In fact, they are a bit like the lion or tiger equivalent of the invertebrate world. Except, that is, that the vision of jumping spiders is in many ways far better than that of a big cat. They have an array of large and small eyes on their head, eight in total. The smaller side-facing eyes play an important role in tasks such as detecting movement, before the spider orientates to look face-on at whatever grabbed its attention. Their big forward-facing primary or principal eyes offer excellent ability to see fine details (acuity), and come with a range of tricks. For example, while the eyes (or lenses specifically) are fixed in the head and cannot move to look around (unlike human eyes, for instance, which can move in the sockets), the retina at the back of the spider eyes can move instead, allowing the spider to shift its gaze. This is equivalent to us looking around by moving only the backs of our eyes where the photoreceptors are. Studies since the late 1960s and '70s have indicated that many jumping spiders likely have excellent colour vision too, and they use a variety of sensory cells and filtering structures to achieve this. A number of species have cells that are sensitive to UV light, and sensitivity to UV light is seemingly very widespread in jumping spiders.

Much of the research aimed at understanding the role of UV light in jumping spider behaviour has been undertaken by Daiqin Li and Matthew Lim at the National University of Singapore. In 2005, Li and Lim reported experiments where they showed that one spider, *Portia labiata*, a particularly ferocious predator of other spiders, uses UV information in deciding whether to invade the webs of its prey.[5] The spider it attacks, *Argiope versicolor*, similarly to many other web builders, 'decorates' its webs with sections of silk. The function of these silk structures has long been investigated across spiders and has been a matter of much debate, but they frequently reflect UV light, including those of *A. versicolor*. Silk decorations often lure prey to the webs, particularly flying insects that can see and are drawn to UV colours. When Li and Lim gave *Portia* a choice of whether to approach webs with UV-reflecting structures versus those without (by removing the UV from the illuminating light), *Portia* preferred to move towards the webs with UV reflectance. It is hard to say why exactly this happened, but the likelihood is that the UV markings act as a cue that *Portia* can use to identify known or suitable prey species, or the presence of an active web.

Lim and Li then turned their attention to the role of UV vision and coloration in spider ornaments and mating behaviour, studying the coloration of different body parts of the tropical jumping spider *Cosmophasis umbratica*, and difference in appearance between the males and females.[6] These spiders are highly ornate, with beautiful iridescent body markings whose coloration stems from the physical arrangement of body scales (Plate 2). Lim and Li found that the body areas used by males in courtship and in displays to rivals always reflected UV light, whereas equivalent areas in females (and juveniles) did not. Thus, the spider is what we call sexually dimorphic—males and females look different. This is common in the natural world, frequently found for example in the plumage of birds, with males usually gaudier. Here, Lim and Li showed that the same was true of the UV coloration of this spider.

Lim and Li's work showed that UV colours are likely to be important in mate choice in these spiders, and also in communication between rivals. Females showed more interest in males when UV light was present, so that the UV-reflecting features show up. And when UV light was removed, antagonistic behavioural displays between males were affected. In fact, males even started to make courtship displays towards rival males (and to their own mirror reflection). This makes sense: since females lack UV reflectance, the males seem to respond as if faced with a female. As such, the presence of UV-reflecting colour patches is important in guiding the appropriate response of males towards potential rivals or mates. Moreover, the male UV coloration in C. umbratica depends on their condition. When deprived of food, the markings on their abdomen become less striking. This means that, potentially, females could be judging males on their condition or quality, how good they might be at finding plenty of food or being a good competitor against rivals.

Most of the work on jumping spiders and other animals with UV vision has been focused on so-called 'near UV' or UV-A light, the wavelengths spanning 315–400 nm, just beyond the blue end of the visible. Many UV-reflecting structures in the natural world occupy this part of the spectrum, and it is widely thought to be the key region involved with UV vision in nature. The higher energy 'far UV' or UV-B light, between 280–315 nm, is less abundant, and considered to be damaging to the eyes of animals sensitive to it. The lenses of many animals' eyes cut out UV-B before it reaches the photoreceptor cells. When animals can detect UV-B, they seem to use it for avoiding exposure to its damaging effects entirely. However, some jumping spiders use UV-B light in communication.

In the highly ornate spider, *Phintella vittata*, from China—a tiny creature, just 4 mm long—both males and females possess markings with reflectance centred specifically on the UV-B range.[7] There is also much variation in reflectance among males, including among populations from different locations, as well as differences between the sexes. Although females have UV-B markings too, the spiders are dimorphic in the amount of UV light they reflect. When presented with potential mates, females prefer males that are shown in light that includes UV-B wavelengths over males shown without UV-B. Just how common UV-B markings and vision is remains to be seen, but visual sensitivity into this region appears to be widespread across jumping spider species, with its occurrence possibly linked to light conditions in different habitats. It is not yet clear what UV-B offers over simply using UV-A, but one possibility, which we will return to shortly, is that if many animals cannot see UV-B light, then it may afford a private channel of communication hidden from the prying eyes of other species, including predators.

In these jumping spiders, UV reflectance is not the end of the story. Many are wonderfully colourful in other parts of the spectrum too, and they have a further feature that can be important in guiding their behaviour: fluorescence. When light of one part of the spectrum is absorbed and then re-emitted as light of longer wavelengths, it is said to fluoresce. Often this involves UV or shortwave 'blue' light being absorbed and re-emitted as green, yellow, or even red. Fluorescence formed the basis of a lot of 1980s and '90s disco styles, as well as various novelty paints, where, under dim conditions with a 'black light' emitting only shorter wavelengths invisible to the human eye, some paints, dyes, and clothing would glow in bright colours in the relative darkness. In nature, the phenomenon is surprisingly common, occurring in both animals and plants, though how often it has any adaptive function is debatable (more on that shortly).

One species of spider that fluoresces and uses this in communication is *C. umbratica*. We know already that male *C. umbratica* have UV reflectance but females do not. Female body structures absorb UV light, but their palps in particular re-emit this as blue-green fluorescent colours. Males on the other hand lack fluorescence and just reflect the UV light (Plate 3). Studies have shown that females no longer court males when the males' UV colours are not visible, whereas males no longer court females when the females' fluorescence is blocked.[8] Although there has been little other work on spider fluorescence, the phenomenon has been shown to be common across a wide range of spider species. The exact nature of the fluorescence, and the colours produced, varies among species and can be quite striking, but the basis generally seems to be that the haemolymph ('blood') of many spiders contains fluorescent chemicals called fluorophores. In species in which fluorescence is especially pronounced, these fluorophores are moved into the cuticle of the body structure where more light can reach, activating them. Fluorescence in spiders seems to have evolved several times independently, which further suggests that it has a function in other spider species too, but without more work we cannot be sure.

While it may seem exotic and exciting, we have known for a long time that many animals can perceive UV light. Sir John Lubbock, a friend of Darwin, showed in experiments published in 1881 that ants could see and respond to UV light.[9] His work was partly in response to common opinion at the time that all animals see the same light and colours. Lubbock, and others working about the same time, showed this was clearly not true.

Over the years, the range of animals known to perceive UV light has grown substantially. Does the ability to perceive UV light offer something special, something that seeing other parts of the spectrum could not? This viewpoint was common among biologists in the 1990s and 2000s, and was certainly propagated in no small part by the fact that humans cannot perceive UV light or discriminate it from other parts of the spectrum. It is an attractive idea partly for that reason—UV seems somehow magical to us. The work on jumping spiders certainly shows that UV can be very important to some animals, and that it can sometimes take precedence over other parts of the spectrum. But the evidence that UV vision is special in a wider context is not hugely convincing.

Birds have often been at the centre of this debate, and since the early 1970s and '80s it has been known that many species see UV light. Diurnal birds have excellent colour vision, involving four different types of cone cell. Many species probably have a more refined sense of colour than we do, with UV detection an important element of this. In nature, a diversity of objects reflect UV, from the plumage patches of potential mates, to flowers, seeds, and insects, and so seeing this light is helpful. The question we really need to ask, though, is whether seeing UV offers some advantage to animals over and above seeing other colours. The evidence in birds is mixed. For example, research testing the foraging behaviour of zebra finches (*Taeniopygia guttata*) for variously coloured seeds when different parts of the light spectrum was removed found that removing longer wavelengths of light actually had a greater effect on foraging than removing UV.[10] Other studies found similar results. On the other hand, one of the main arguments for why UV vision and signals in UV may be special to birds is that, while many birds perceive it, their predators might not. And these include various birds of prey, which tend to lack good UV sensitivity. The idea is broadly consistent with evidence that UV markings in the plumage of songbirds may be very visible to other songbirds, but is likely to be less conspicuous to birds of prey like raptors.

Beyond birds and spiders, a handful of other examples show how UV may be especially valuable in communication. In two species of reef damselfish, UV markings play an important role in defending territories and recognizing species. The faces of these damselfish have variable UV markings that we cannot see but they can, and they can use these to distinguish one species from another.[11] As with birds, many of the predators of the damselfish probably can't see UV light, partly because their visual system cannot detect it well, but also because UV light is scattered quickly over short distances, making it a poor source of information over a long range. So, the damselfish may use their UV signals to communicate with one another without attracting the attention of predators.

Ultraviolet vision may also be particularly valuable in some extreme environments, such as in snowy latitudes. Reindeer not only have to cope with freezing temperatures and darkness in the winter, but for much of the year they have to find food hidden in the snow and ice too. They tend to feed on lichen, and some lichen absorbs ultraviolet light, appearing dark, whereas snow reflects UV strongly. In the snowy conditions, the lichen stand out to the reindeer since they can see UV light. The other potential use of UV is for detecting predators. Animals such as wolves have fur that appears dark in UV, even when to human eyes the coat is white. To a reindeer, an approaching wolf might stand out strongly against the snow in UV. The same has also been suggested for some seals that have ultraviolet vision, and may use it for detecting predators like polar bears.

There are times when UV vision and communication signals seem to be particularly valuable over other parts of the spectrum,

yet it's hard to escape the fact that UV has received so much attention in part owing to our fascination with its mysterious nature.
The likelihood is that, overall, UV is rarely more important than
other colours in nature, but each part of the spectrum matters
more on a case-by-case (or species-by-species) basis. If there's
one thing we have learnt about UV vision, it is that UV sensitivity
is extremely common in nature and humans are among the
unusual species in not seeing it.

What about fluorescence? Prior to the work on spiders, other
studies had shown that fluorescence seems to be used by certain
animals in mate choice. One such example concerns a common
pet. Budgerigars have plumage patches on their cheeks that fluoresce under UV light. In fact, if you put a budgie under a black
light (people have), these plumage patches positively glow yellow. Studies at the turn of the millennium showed that male and
female budgies prefer members of the opposite sex more when
fluorescence is present.[12] The fluorescent plumage tends to be
found close to blue colours that reflect UV light. So, the fluorescent patches absorb UV light and reflect yellow, whereas the blue
patches reflect UV-blue light, creating strong visual contrast to
the eyes of the birds.

A lot of work, especially in the past decade, has revealed that
fluorescence is extremely common in nature, occurring from
arthropods to sea turtles. Much of this work has involved people
taking a black light, which emits a large amount of UV and shortwave light, out at night and simply documenting cases of fluorescence. It is especially widespread in the marine environment,
found in animals from corals to fish. Other cases of fluorescence
have long been known and even provided a useful tool—for
example, those wanting to find scorpions have long known that
they fluoresce a bright blue-green at night under a black light.
However, there are two major problems with much of this work

regarding the relevance of fluorescence and whether animals actually see it and make use of it. First, we need to be careful with storytelling: just because fluorescence occurs in a species does not mean that it has a function. The glowing reds and greens found over a coral reef and even on the back of a turtle are undoubtedly beautiful, but they need not necessarily serve any biological purpose. They may simply be a by-product of chemical processes. Second, the above approach is very unnatural. No animal goes around a coral reef at night with a high-powered black light strapped to its head. The light that is inducing the fluorescence is simply not normally there, at least not to that extent. Conversely, under daylight, the intensity of light from the sun will often be so high that it would normally swamp any chance of seeing the fluorescence itself.

Nonetheless, the spider and budgie examples indicate that fluorescence can be valuable under certain conditions. In most cases, it will tend to enhance the visibility of a signal that is already present, such as making the yellow feathers of the budgie brighter and more vivid. Fluorescence should also be more valuable in relatively dark environments and times of day (like dusk or dawn). Here, enough short or UV light is around to induce fluorescence, but it is not so bright that the full sun's light would swamp any fluorescence effects. Accordingly, recent work on fluorescent sharks, for example, shows that two species of deep-sea 'catshark' with green fluorescence have a visual system that may be able to detect some of this under natural conditions in these dark ocean depths, though the potential function of the fluorescence is unclear.[13] In other fish, red fluorescence may be detectable to one another, and individuals of those species do seem to respond to the presence of it in behavioural experiments, including interactions between rival males. Beyond the aquatic realm, there even exist tree frogs with green fluorescence (Plate 4), and scientists have calculated that this may contribute as much as one-third of the total light emitted or reflected from the frog's body under twilight conditions. The frog's visual system may detect these differences in brightness. A remarkable discovery made recently is that pea aphids, which have the ability to discriminate some colours, may use their vision to detect the UV-activated fluorescence of pathogenic bacteria. By doing so, they can avoid getting infected. Returning to the jumping spiders, the fluorescence may be especially important under relatively dark forest canopies with lots of scattered UV light. As with UV, it is easy to be drawn to the 'magical' aspect of fluorescence, but ultimately we will just have to wait to see how important fluorescence is and how widely it is used in nature.

Animals vary in the diversity of colours that they can perceive, and some lack colour vision entirely (Plate 5). Sea lions and certain nocturnal primates have just one type of cone and essentially see the world in shades of black, grey, and white. Other animals do see colour, but a reduced range compared to us. Your average pet dog, for instance, is a dichromat with two types of cone most sensitive to shortwave and mediumwave light. We can never know for sure what colours they see, just as you can never know what colours another person sees, but, in human terms, they essentially perceive blues and yellows. By not having both a 'green' and a 'red' cone, they cannot tell apart red, yellow, and green colours as most humans can. A red ball in the green grass is not particularly visible to a dog, nor would be a green food bowl on a red carpet to cats.

While some animals see fewer colours than us, many probably see far more, starting with those that are tetrachromatic. As the name suggests, these animals are thought to see colour using four receptor types, and included in this group are probably most birds, some fish, reptiles, butterflies, and more besides. A bird such as a blue tit or European robin has cone types dedicated to UV, blue, green, and red light. Not only that, but birds also have deposits of pigments called oil droplets in their photoreceptors, which filter out certain wavelengths of light before they reach the visual pigment. The effect of this is to reduce the overlap in sensitivity to light wavelengths between the cone types. In the process, colour discrimination is enhanced because each cone is more specialized at seeing a specific range of light that the others do not detect.

We cannot readily imagine what the world must look like to a tetrachromatic creature. A trap would be to think of birds seeing a world like ours but with ultraviolet simply 'added in'. In reality, their vision is more complex and interesting. While humans can compare between three cone types, birds can compare between four, and this potentially means a world of far more colours. Imagining this world is a challenge. When we use TV or computer screens to show film or images it is based on the principle of having three light types in the display, like the phosphors of old TVs, emitting red, green, and blue light. We can mix different amounts of these light types to simulate how a scene would have looked in reality by making light mixes that stimulate the cells in our vision in much the same way that our eyes would be stimulated as if seeing the real thing. It isn't perfect, not least because each display differs, and so we sometimes see colours that seem a bit odd or inaccurate, though with modern TVs this is less common. The problem is that no technology exists to display the range of colours that tetrachromatic animals can see—we would need four-layer images or four phosphors, and even then, we only have three cone cells to view them with anyway. We can add extra information into images from other parts of the spectrum, but this really just enhances one of the colour types we see already rather than adding more colours.

One way to think how colour vision works is to imagine colours occupying regions of a 'colour space'. In this way, trichromatic vision can be represented as a triangle. For humans, the three points of the triangle correspond to the three cone types, red, green, and blue, and the centre of the triangle is where all three cones have the same stimulation (that is, there is no colour but just some shade of black, white, or grey). Various points within the triangle can represent the range of colours that in theory could be seen. Any given colour we see with our visual system could be plotted in this space, based on the relative stimulation of each cone. A colour that we would see as rich in red would fall near the red tip. A colour that's purple would fall in-between the red and blue, and away from the green tip. Now imagine you add an extra receptor for seeing colour, and we now have to change the 2D triangle into a 3D space, a tetrahedron, with four tips and a range of potential colours in this 3D volume. So, adding an extra receptor type to colour vision does not just add one more type of colour, it potentially adds a whole extra dimension of colour space (Figure 11).

For the vast majority of animals, colour spaces are broadly how scientists tend to conceptualize the range of colours that they might see. However, one of the most remarkable and unusual animals in nature, the mantis shrimp, appears to have thrown out the rule book when it comes to seeing colour. Mantis shrimp, or more accurately stomatopod crustaceans since they are not actually mantis, are highly dependent on vision. Comprising over 400 species, they first began to arise some 400 million years ago, and now exist in a myriad colour forms. If anything on Earth looks truly alien, it is mantis shrimp. They are

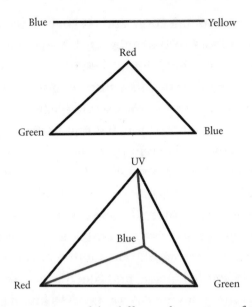

Figure 11. Representations of the different dimensions of colour space that some animals have. Dichromatic animals (top) have just two receptors for colour vision, and so we can think of the range of colours they see as falling along a single line, with the central point being grey. At each end would be sensations of a given colour type (e.g. blue or yellow). Trichromatic animals (middle), including humans, see colours using three receptors, and we can represent this by considering colours falling within a triangle. To illustrate the colour diversity potentially seen by a tetrachromatic animal, with four receptors, we need to consider how colours would fall within a 3D tetrahedral space (bottom).

ferocious predators, fast to strike at prey and known to either bash through prey armour at up to 50 mph or spear them with their raptorial appendages. Being fiercely competitive and defensive of their home burrows, they also display a range of social behaviours. The diverse marine environments in which they live, from coral reefs to open rocky sea beds, often with relatively bright conditions, enables mantis shrimp to communicate with a wonderful diversity of colour patterns. In fact, some mantis shrimp even have fluorescent yellow patches on areas of their body used for communication, and this may enhance the vividness of their colours to one another.[14] Aside from being very vibrant in colour, it is hard to miss one of their most prominent features: their huge eyes that they move and twist around on the ends of stalks as they inspect their world (Plate 6).

Mantis shrimp are bizarre in the number of receptors they have, each sensitive to different parts of the light spectrum. The compound eyes of the mantis shrimp are made of arrays of cells called ommatidia, each with a lens focusing light onto receptor cells below. There are three main areas of the eye: a central 'midband' of six rows of cells running across the eye, and two peripheral regions of cells above and below this. It is easy to see the midband in a close-up image, and much of the work on these charismatic animals has focused on the cells found there.

Studies seeking to uncover the mysteries of mantis shrimp vision include much work by Tom Cronin at the University of Maryland and Justin Marshall at the University of Queensland.[15] In 1988, Marshall found that cells in the midband are very diverse and perhaps enable vision of both colour and polarization (we'll come to polarization shortly). A year later, Cronin and Marshall published studies analysing the cells in the eye of two species, *Pseudosquilla ciliata* and *Gonodactylus oerstedii*, again focusing on photoreceptors in the midband. Here, owing to a range of different visual pigments sensitive to different wavelengths of light, and some elaborate filtering mechanisms, the animals effectively create ten or more types of photoreceptor detecting light in different parts of the spectrum. Fast forward a decade and research on another species, *Neogonodactylus oerstedii*, showed that these mantis shrimp have no fewer than four types of receptor (three in the midband alone) dedicated just to seeing different regions of UV light.

Since then, the number of receptors for sampling across the spectrum has been shown in some species to be as high as twelve or even sixteen, including up to six types tuned to UV. All in all, the mantis shrimp can therefore discriminate light widely across the spectrum from 300 to 720 nm, based on cells in four of the six midband rows, with a great many receptors sampling specific regions of the spectrum. On first assumptions, this sounds like an amazing visual system for the mantis to have, but here comes the conundrum. When vision scientists have done some clever maths to analyse a huge range of reflectance spectra from natural scenes, representing anything from vegetation and other animals through to rocks, they find that having just three to four different types of receptor should be more than sufficient to interpret pretty much the entire range of colour information an animal is likely to encounter. If we take our understanding of how colour perception in most animals works, you simply should not need to be more than a tri- or tetrachromatic animal. In fact, having more receptor types for colour vision may be a waste because in order to pack more receptor types into the eye, you lose some ability to compare between each type.

Colour variation and types in nature is limited, and, sure enough, the majority of animals do indeed have no more than three or four receptors for colour discrimination, and often fewer. So the mantis shrimp represents a puzzle. A natural assumption had been that mantis shrimp must have an astounding sense of colour perception and discrimination. Surprisingly, then, recent work has shown the opposite: their ability to discriminate between colours is actually not that great. In fact, it seems to be worse than the abilities of humans and various other animals. Hanne Thoen at the University of Queensland and colleagues tested how good one species of mantis shrimp (*Haptosquilla trispinosa*) is at discriminating between light of different wavelengths.[16] They trained individuals to tell the difference between the lights by giving them food rewards when they got it right, and found that the animal's performance was much worse than would be expected had they been using a type of colour vision similar to most other creatures. That is, if they discriminated between colours based on comparing the outputs of the receptors in a type of opponent colour system then they should have done much better. Instead, it seems that the mantis shrimp scan across the visual range from UV to longwave and take regular samples across this spectrum with each receptor, but apparently they do not compare the outputs of those. Based on which receptor is active they may recognize a limited number of colours. This is broadly similar to how remote sensing satellites work, whereby samples taken across different wavebands are compared to a sort of look-up table.

The explanation for mantis shrimp colour processing remains an idea yet to be fully tested. Future work will have to establish exactly how their vision functions. Beyond this, the other puzzle is in knowing why mantis seem to do things so differently to practically all other animals tested so far. Again, we do not yet understand why, but the answer may lie in the speed of processing light information their system affords. Rather than having nerve systems that need to compare receptor types, the output from each mantis shrimp photoreceptor could go straight to the brain and activate a dedicated place for processing information from that receptor. Thus, if 'receptor 6' fires, it tells the brain that a specific colour, or part of the spectrum at least, is present. Mantis may recognize a limited range of colours, but are extremely fast at doing so. Their brain is much smaller than ours, yet they are processing information from far more receptor types than we have, and it may simply be beyond their means to compare all that information directly like we do. In their world, where speed is key, rapidity may be what matters in processing colour for mantis shrimp.

There is more to the mantis shrimp than an unusual and still somewhat mysterious way of seeing colour. They also have other special powers that are well beyond human abilities and indeed those of all other animals known. Mantis have an outstanding ability to see and use different types of polarized light. Polarized light is not an easy thing to get your head around, so we'll try to keep it simple here.

Light acts both as a wave and as a stream of particles (photons). As an electromagnetic wave, it consists of electric and magnetic fields which vibrate at right angles to each other and to the direction in which the wave is travelling. The direction of vibration of the electric field is represented by the e-vector. When light has waves with e-vectors that change randomly over time, on average, then there is no predominant e-vector direction and the light is unpolarized. However, when light is made up predominantly of the same e-vectors, as happens for instance when it passes through a polarizer, which only lets through e-vectors aligned in a particular direction, it is said to be plane polarized (Figure 12). Light coming direct to us from the sun is unpolarized, but when it gets reflected or scattered, it can become polarized. This is why polarized sunglasses reduce glare—they can stop the reflected, polarized component of light.

In real visual systems, when the visual pigment in a photoreceptor cell is activated the chances of this occurring are greater when the receptor molecule is aligned with the e-vector of light, meaning that the pigment molecules are naturally sensitive to polarization. However, in humans and other animals that do not perceive polarized light, the visual pigments in the membranes of the photoreceptor cells sit in different orientations, meaning

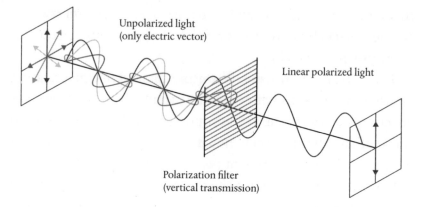

Unpolarized light
(only electric vector)

Linear polarized light

Polarization filter
(vertical transmission)

Figure 12. Representation of polarized light, showing how unpolarized light has an electric field that vibrates in a variety of directions over time. By contrast, polarized light comprises waves that predominantly vibrate with a particular direction.

that the cell responds to many e-vectors rather than being biased in one direction. By contrast, in many invertebrates the visual pigment molecules are held in specific orientations, meaning that the cells detect polarized light. Varying the orientation of the molecules in different cells makes classes of receptor able to detect different polarization angles, which can be compared, potentially in a manner a bit like a colour opponent channel.

Using polarization imaging methods, Marshall and others found that some mantis shrimp produce patterns of polarization in light reflected off their bodies, including their antennae, suggesting they may use these signals in communication.[17] They also showed that the animals could learn the orientation of polarized light, and discriminate between orientations independently of other visual cues, such as brightness or colour. Comparing polarization of light based on having types of receptor in which the pigment molecules are arranged in different orientations is referred to as 'static' polarization vision. Correspondingly, there exist cells in the mantis shrimp eye (in rows 5–6 of the midband), particularly centred on the UV part of the spectrum, that are arranged in such a way as to be able to compare different angles of linearly polarized light.

Static polarization vision is in contrast to 'serial' polarization vision, in which cells with the same arrangements of molecules are compared over time as the animal moves its eyes or head. In species that can rotate their eyes, the orientation of the cells can be changed, allowing comparisons over time to be made using polarization patterns, or for the animal to enhance its ability to see polarization information. Mantis shrimp do just this, and their wonderful ability to rotate each eye enables them to align the photoreceptors to the angle of polarization in a way that best enhances the contrast of polarization information on an object and its background. As such, the mantis shrimp combine specific behaviours and body movements to enhance their vision even further.

Somewhat like UV, polarization is actually very common in nature even though we are largely blind to it. Bees or ants use patterns of polarization in the sky for navigation, other insects use it for finding sources of water, and some animals beyond mantis shrimp, such as cuttlefish and butterflies, communicate with polarization patterns on their bodies, including for mate choice. However, another more exotic type of polarized light can exist: circular polarization. When a wave of light travels, the e-vector can rotate, rather like a corkscrew, and it can do this in either a clockwise or an anti-clockwise direction (right- and left-handed, circularly polarized light). Mantis shrimp have been shown to be able to tell the difference between types of circularly polarized light.[18] Measurements of their eyes have revealed cells that are arranged in a manner and structure that would allow them to do this. In addition, crucial evidence has included tests showing that individuals can be trained to respond to either left or right circular polarization. At least three species of mantis shrimp have also been shown to have patterns of circular polarization that can differ between males and females. Further research confirmed that one species, *Gonodactylaceus falcatus*, not only has patterns of circularly polarized light, but they can detect these patterns and use them in their behaviour. Crucially, mantis shrimp avoided burrows when they detected the presence of circularly polarized light, a sign that they may be occupied by other individuals. Mantis shrimp are highly aggressive, so if they can avoid coming into contact with rivals by detecting each other's presence then they can avoid potentially costly encounters.

Michael Land, an expert on vision from the University of Sussex, writing in 2008 about the initial discovery of circular polarization vision, began his report: 'I had intended to get through life without having to understand circularly polarised light. That hope has now been dashed by a flamboyant crustacean...'.[19] Mantis shrimp have certainly made vision biologists reflect very hard about what they think they know! Circular polarization vision and signals have been suggested in one or two other groups of animals as well, but the evidence is contested. From what we know at present, beyond mantis shrimp this is an ability that few if any other animals in nature can get close to. Yet again, why mantis shrimp need such capability is unclear, but as with other phenomena, being able to see circularly polarized light may afford them another private communication channel.

A key aspect of vision (and senses more widely) that we are only starting to realize is just how plastic it is, and how it can vary from one individual to the next. Or even how it can change within an animal's own lifetime. Far from being fixed at birth, the senses can change and respond to the world. Many species of mantis shrimp have visual sensitivity tuned to the light spectrum where they live. As you descend into a clear ocean, the water selectively removes longer wavelengths of light first, so that not only does the environment become darker with increasing depth, it also becomes blue-shifted. In species of mantis shrimp that live deeper down, some of the cells in their eyes are less sensitive to long wavelengths and more sensitive to shortwave light. This makes sense because for their vision to function in the deeper water it needs to be sensitive to the light available. It is a similar story in many other marine organisms, including very deep-water species, since most bioluminescence is also blue-green. This tuning of vision to the light environment occurs even within the same species. In one mantis shrimp found in Australia, *H. trispinosa*, individuals found in deeper water have vision more sensitive to blue light. When individuals are raised under blue-shifted light in the lab it causes them to shift their vision during development to see shortwave light better than those reared under white light.[20] This is important because it shows that the way an individual's vision works is affected by the environment.

Other factors can also have an important influence on how vision is modified. Guppies are beautiful in coloration, and vision is crucial in their mating behaviour. If you supplement the diet of guppies with orange-red pigments called carotenoids, then those individuals grow up with a visual system that seems to use longer wavelengths of light more, and potentially enables them to see red better. Carotenoids are something that many animals get from their diet and they are an important part of the pigments used in vision. The old saying about eating lots of carrots giving you good night-time vision might not be precisely true, but doing so might give you better colour vision, since carotenoids are abundant in the vegetable.

Changes in vision can also be in sync with the season. Reindeer must find food in the snow, but they also face extreme changes in light conditions during the year: dark and snowy in the winter; almost permanent sunlight in the summer. That is a considerable challenge for their visual system and they have a curious way of dealing with this. At the back of their eyes is a structure like a mirror that reflects light back and enhances vision in low light conditions: a 'cat's eye' (or *tapetum lucidum*). In summer, this is yellow in colour, whereas in winter it goes blue, and presumably allows the reindeer to gather light from parts of the spectrum that are most abundant at different seasons, including UV. While it is tempting to think of vision, and indeed other senses, as fixed, it is probably much more variable and changeable within the lives of animals than we tend to appreciate.

Before finishing our tour of vision, we should consider how animals use information gathered by their visual systems to behave appropriately. An animal that has long been used to test this is the toad. Toads like to eat small invertebrate prey like beetles and worms, and when they see a suitable meal they characteristically 'snap' their mouth towards them. The question is how toads recognize prey from non-prey, and this has been a major area of research for Jörg-Peter Ewert, a German neurophysiologist at the University of Kassel, and colleagues.

One of the key features of many of the invertebrates that toads eat is their elongated body shape. A worm has a long body, and many beetles are longer than they are wide. Also, they move in the direction of this elongation. These two features combined mean that there are simple key characteristics that could be used to define suitable prey: something that is long and moving in the same direction. Studies in the 1960s and '70s onwards showed that common toads (*Bufo bufo*) and other *Bufo* species show characteristic snapping behaviour towards a simple small dark rectangle moving in the appropriate direction (Figure 13).[21] Yet they

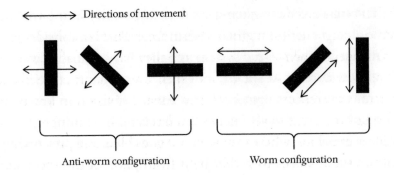

Figure 13. During prey capture, toads show a very characteristic snapping attack response towards potential moving prey that are elongated in body shape, such as a worm. So stereotyped is this response that toads can be induced to orientate towards and attack a simple rectangle moving in the direction of the long axis ('worm'). By contrast, they show little response, and even aversion, to rectangles that move in other directions ('anti-worm').

ignore or even avoid the same rectangle when it moves in a direction perpendicular to the long axis. For example, a vertical rectangle moving horizontally would not be attacked.

Various similar experiments showed that the toad has a rapid response to 'worm-like' stimuli but avoids equivalent stimuli when the aspects of movement and shape disagree. These experiments were simple but elegant. For example, one approach was to place a toad in the centre of a transparent cylinder and then mechanically move a dark cardboard rectangle around it in a circle. When the toad was interested, it jumped around in a circle trying to follow and snap at a shape. Similar effects can be demonstrated when presenting toads with computer-generated rectangles moving across a screen in front of it. In essence, the vision of the toad acts as what is often called a 'feature detector', producing a quick and appropriate response to a set of criteria that reliably encode objects of particular importance—here, food.

The toad's visual system acts as a feature detector based on several stages of visual processing. In the eyes are so-called retinal ganglion cells, which process information that is initially detected by the photoreceptors (e.g. cone cells). The ganglion cells link the outputs of sets of receptors to create something called 'receptive fields'. We have already encountered receptive fields in the context of hearing, but they encode particular patterns of information, which in this case relates to object size, orientation, and movement. For example, some receptive fields will only be activated by objects of a specific size, equivalent to that of a suitable prey type. As objects move across the visual space of the toad, receptive fields become activated and pass this information to the midbrain. Some of these ganglion cells (so-called R2 classes) are very good at encoding worm-like stimuli.[22]

The information from the ganglion cells is processed in the midbrain. Here, there are also neurons that produce receptive fields, and different ones are active depending on combinations of information sent from the eyes. The location of objects is represented as a spatial map, whereby particular cells fire depending on the actual location of the object in space in the real world. When a relevant receptive field in the brain is active, the toad can then turn to face the right direction. Cells in the toad's visual processing system respond only to activation that is occurring in a small area of space in the outside world. That is, they are not activated by large-scale movement patterns across the eye that would occur when the toad was moving through the undergrowth or vegetation was being blown around by wind. Only movement in certain small areas denote the presence of a prey item.

More specifically, information from the ganglion cells passes to a set of neurons in the optic tectum called the T5 neurons, and some of these are activated by worm-like stimuli, identifying the object as prey. This region then produces the prey capture behaviour. In contrast, information from receptive fields that relates to anti-worm stimuli, and other inhibitory things such as a looming object like a predator, pass to another area (the thalamic-pretectal region) and so-called TH3 neurons. These have inhibitory projections to the T5 region, preventing capture behaviour when it would be inappropriate or even dangerous. The combined effects of the TH3 and T5 neurons ultimately lead to either attack or avoidance behaviour.

The toad's visual processing shows how tasks of special importance to an animal often have dedicated pathways and areas of the brain for encoding and processing key aspects of the stimulus. Prey capture in toads is both of critical importance (they have to eat to survive) and suitably stereotyped, so that some relatively simple pathways from the eyes to the brain, and then on to the motor systems that control behaviour, can produce rapid relevant responses before the prey can escape. The whole thing is relatively fixed or 'hard-wired', which is why toads show such strong reactions to paper rectangles being moved in front of them.

Vision in animals is wonderfully diverse, and part of this is because animal eyes can potentially encode so many different types of information from the surroundings. This includes different regions of the light spectrum, spatial information such as patterns, temporal changes such as the movement of objects, polarized light, and more besides. The diversity in types of information to which different animals respond depends on the animal's ecology, where it lives, and tasks of particular importance. The study of vision has revealed much about how sensory information is processed, and given rise to plenty of debate and mystery too. It has recently also played an important role in appreciating just how plastic and adjustable aspects of the senses can be when the environment can change. In Chapter 4 we turn to a type of sense that is equally important to many species, but also one that we lack entirely: the electric sense.

CHAPTER

4

ELECTRIC
ATTRACTION

We have already seen that some animals have astonishing powers of hearing and vision, far beyond our own. But there are some senses that we lack entirely, one of which is the ability to detect and interpret electric information. It is hard to conceive of sensory worlds that are so completely beyond our means; to imagine what it must be like to perceive a type of information to which we are blind. The challenges in appreciating and testing how some animals use electricity in almost every aspect of their daily lives are considerable. Nonetheless, substantial progress has been made and we know that the electric world of many species is as rich, exciting, and complex as that of any other sense.

In the 1950s and '60s, a growing body of work had uncovered clear evidence that some fish could produce electricity in order to detect objects and communicate with one another. That research had also begun to show the presence of special receptors for detecting electric information, and even how those receptors work. However, it was also apparent that some species, especially sharks and rays, had what seemed to be electroreceptors but seemingly no way of producing electricity themselves.

The question of how, and for what, these sharks and rays use their receptors was yet to be solved.

Sharks are revered (and often unfairly feared) for their sophisticated ability to hunt. How exactly they locate their prey depends on the species and circumstances, but like any predator they use an array of sensory information. From a longer distance, many sharks use odour cues, being famously sensitive to tiny amounts of blood in the water. Closer up, the silhouette of an animal seen against the water surface, the movement patterns it makes, or even the sparkles of light that reflect from the bodies of many fish can all be important visual cues for targeting dinner. For many sharks, however, this is not enough. They also rely on additional, highly sensitive information in the form of electricity. While most animals do not have specific cells or organs to produce electricity, certain daily natural biological processes, such as moving or breathing, cause tiny amounts of electricity to be created, and in principle a predator could use this to its advantage in locating hidden targets.

The clearest demonstration of why sharks and rays had electroreceptors and how they used them was revealed in a study in 1971 by Adrianus Kalmijn from the University of Utrecht in the Netherlands.[1] He presented a small-spotted catshark (*Scyliorhinus canicula*) with a hidden live prey, a flatfish (*Pleuronectes platessa*), and showed that the catshark could detect the prey using electric cues alone. The experiments were wonderful in both their power of systematically arriving at the final answer and in being elegantly simple in nature (Figure 14). In the first part of the experiment, Kalmijn presented the catshark with a flatfish that was completely hidden in the sand, to eliminate visual cues. The shark had no trouble in finding it. Next, he placed an agar chamber around the hidden fish, to remove chemical and mechanical, but not electrical cues as well. Again, the prey was easily found

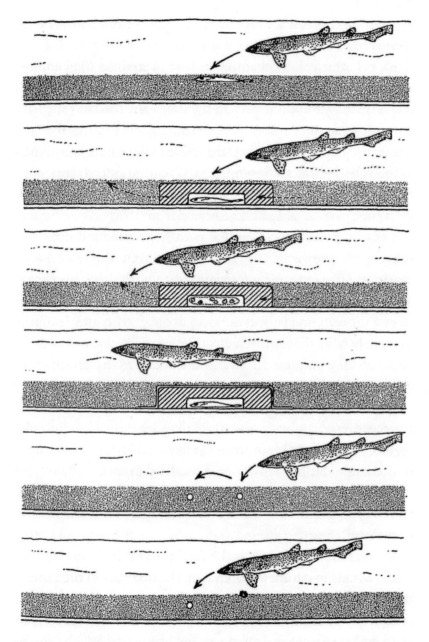

Figure 14. Kalmijn's outline of his 1971 experiments that demonstrated how catsharks use electricity to locate a flatfish hidden in the substrate. When the flatfish was buried, and even buried surrounded by a chamber to prevent visual and chemical cues, the catshark had no trouble finding the prey. However, when the chamber was electrically insulated the shark swam past. In contrast, it would readily attack buried electrodes, even in preference to a piece of fish bait on the substrate nearby.

by the shark. When Kalmijn pumped water from the chamber containing the flatfish, so that it flowed out elsewhere, the cat-shark still attacked the chamber instead of targeting the outflow. On the other hand, when it was electrically insulated, the shark could no longer find the prey in the chamber. This was strongly indicative that the shark needed electric information for the final phase of prey location, yet the clinching proof came when Kalmijn buried electrodes in the sand that produced weak electric signals; these again attracted the catshark's interest. In fact, the shark even attacked the electrodes preferentially over fish bait placed nearby, showing that electric information can override chemical or visual cues in close proximity. Therefore, while sharks often use chemical information to track down prey from a long distance, many species enlist their electric sense to determine the prey's precise location and to direct their attacks. Their extremely refined, highly sensitive electric sense allows sharks to detect electric cues produced by prey bodies as low as 0.01 mV/cm, and perhaps even down to 15 billionths of a volt.

The shark's electric sense comes down to a series of cells located around the body, but especially concentrated in the head region. Were you to inspect the underside of a shark you might notice lots of little holes or pores on the skin surface. These are filled with jelly and are in some ways similar in nature to the hair cells of the lateral line that fish use to detect the mechanical movement of water around them. We call these cells electro-receptors, and in cartilaginous fish like sharks and rays they are found in clusters in a feature known as ampullae of Lorenzini—named after Stefano Lorenzini, who described the pores on the bodies of sharks and rays in 1678 (Figure 15). The cells respond to changes in the electrical charge of the jelly inside. Much later, in the nineteenth century the similarities to the sensory receptors found on the lateral line of fish was realized; however, the

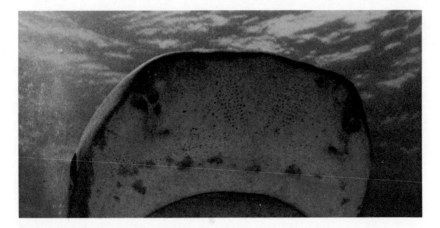

Figure 15. On the head of a tiger shark can be seen the ampullae of Lorenzini, the pores containing sensory cells that detect electric information.

ampullae detect not water movement but the tiny electric fields created by the bodies of prey animals in water. So even if a shark is drawn to an area by its sense of smell and the location in which the prey animal is hidden, it can still rely on its electric sense to home in on the invisible target.

Electric information can also be used for defence against predators. Some sharks lay their eggs in so-called 'mermaid's purses'; occasionally, we can find these washed up on beaches or in tidal pools. The red-brown pouches contain the developing shark embryo, and, while inside, the baby shark begins to move around as it grows. Shifting around, however, carries a risk because predators might detect the movement of the shark inside and discover a tasty bite size meal. Researchers have shown that some sharks, specifically the brown-banded bamboo shark (*Chiloscyllium punctatum*), guard against this risk using their electric sense, which develops while they grow.[2] In experiments, when the shark embryo is exposed to a weak electric field, similar to that of a nearby predator, they freeze and stop moving, pausing even their gill movements. They are not alone in this response, since cuttlefish also seem to possess a similar ability to freeze when they detect electric information. So, although it is normally used for prey detection, the electric sense can sometimes be used in defence too.

Detecting minute amounts of electricity is an ability that occurs in a surprising range of animals. Normally, it is limited to aquatic species, because water is a much more effective conductor of electricity than air, and fish prevail when it comes to such capabilities. Exceptions do exist though, including animals as diverse as the echidna, and potentially some dolphins and crayfish, though even more unusual examples have recently been found that we will deal with later. But let's look first at the masters of the electric sense: the electric fish.

The detection of very low levels of electricity from prey by sharks and rays is extremely impressive, but some fish are capable of much more. Rather than simply perceiving electricity around them, these 'weakly electric fish' emit their own electric signals into the environment—they are 'electrogenic'—and measure changes in the field when it interacts with their surroundings. By doing so, they can gain crucial information about features and objects in their environment. A power like this is analogous to the echolocation abilities of bats, and is another so-called 'active sense'. In some strongly electric fish, including torpedo rays, certain catfish, and electric eels, the amount of electricity produced is so great (up to and over 600 volts) that it can be used to stun prey or act as a defence. However, a far greater number of fish produce weaker electric signals (typically less than 1 volt) that they use for navigation, detecting objects and prey items, and even communicating with one another. Their abilities are truly impressive, enabling individuals to, for example, discern the size, texture, shape, and even the distance to objects around them in the water. Species that are nocturnal or live in dark conditions, or where the water is very murky and vision is of less value, tend to be those that use electricity. As far as absolute numbers of fish species go, these are not especially common; about 350 species are electrogenic from the 30,000 or so fish species overall. But the ability is prevalent in a number of groups.

Knowledge that some fish produce electricity is far from being a modern-day discovery. Indeed, fish capable of producing large amounts of electricity were historically prescribed for treating a variety of health issues in various parts of the world. The literally shocking properties of some fish were even known to ancient Greek and Roman philosophers. Investigations of the anatomy of these fish, including of torpedo rays, largely began in the 1600s, with attempts to discover the source of their powers. Only from the mid-1700s, however, after the development of electrical theory, did people begin to appreciate the similarity of shocks produced by fish with that of early devices for storing and producing electricity. And it was noted that the shocks from animals such as the electric eel could occur through water when the person was some distance away from the fish.[3] A series of experiments by John Walsh, who was not only a fellow of the Royal Society, but also a member of Parliament, showed the first clear evidence of electricity production by animals. In one of his most important experiments in 1776, he demonstrated how, with a combination of wires and electric circuits, the shock of an electric eel could create a visible spark in a dark room. This elaborate demonstration showed beyond doubt that some fish were capable of generating electricity.

Nevertheless, the actual functions of electricity, especially beyond defence, were less clear. Scientific understanding increased considerably with studies from the 1950s onwards, including classic research by Hans Lissmann of Cambridge University. In 1951, Lissmann published a study of the fish *Gymnarchus niloticus*, showing that it had an exceptional ability to avoid objects in a tank while swimming backwards.[4] It could even navigate through tight spaces, such as between rocks. By placing electrodes in the tank he could record the presence of continuous electrical activity produced by the fish. Moreover, the fish would actively respond to the presence of electric information, or to objects that conducted electricity (such as copper wire) in the tank, with either avoidance or aggressive behaviour.

A few years later, Lissmann revealed further discoveries from his studies of various species of electric fish. He showed that species differ in key features of their electric discharges, including their frequency and amplitude, and that they use their electric sense in a strikingly wide range of tasks, from navigation through to prey capture and communication. A major feature is that they can not only detect the electricity made by other fish, but also measure distortions to the electric field that they themselves produce. Research over the following decades showed that the range of the electric fields produced by the fish is quite short, detecting objects within about 10-15 cm. Nonetheless, some groups of fish have highly refined abilities to detect the size, shape, texture, material, and distance from an object in their environment, giving them valuable information with which to interact with the world.

Electricity production has evolved a number of times but two main groups of weakly electric fish exist, which we will come to shortly. In these animals, the electric organ evolved from modified muscle and nerve cells. It comprises special cells called electrocytes, each of which produces small amounts of electricity. These are arranged in rows, in series, so that the combined current and voltage is summed, a bit like batteries connected in series, to produce a much larger output. When the fish starts an electric signal, called an 'electric organ discharge' (EOD), many electrocytes all produce a synchronized response. Each cell creates an action potential, which like any nerve cell impulse involves an influx of negative sodium ions over the cell membrane. The combined effect of many electrocytes firing together is that the skin of the fish becomes electrically polarized, resulting in the creation of an electric field. By changing the location along its body where groups of electrocytes fire, the fish can alter features of the electric field in time and space, and measure how the field interacts with objects around it. For example, the black ghost knifefish (*Apteronotus albifrons*) generates electric field lines that arc out around its body and into the surrounding water. This knifefish is native to South America, including in the Amazon River, where it hunts for small invertebrates and navigates through the river systems. It is a rather striking fish, black in colour with two white rings on its tail, and the common name arises from an old belief among some indigenous South Americans that ghosts had possessed the fish. As the discharge arcs around it, features of the environment cause distortions to the electric field: some objects act as conductors and attract or 'pull together' the electric field lines, whereas non-conducting objects repel or 'push' the field lines apart. The location of different types of distortion around the body allows the fish to determine the presence and features of objects. Indeed, the animal will often swim in a manner that brings the most sensitive parts of its electric field (highest current density) into the region of unknown objects so they can examine them more closely.[5]

Lissmann observed that the electric sense of fish must have evolved several times independently, since it is found in unrelated species and the electric organs occur in different parts of the body in these groups. We can think of this as similar to ears occurring in different parts of the bodies of many distantly related insects, from their heads to their legs, all with the same general function of detecting sound. Current understanding is that the evolution of an electric organ has apparently occurred seven or eight times independently in different groups of fish. In fact, in the *Origin of Species*, Darwin suggested the electric organ as a potential example of convergent evolution.

Of the weakly electric fish, many come from two independent evolutionary radiations (rapid division into many species): the Gymnotiformes or electric knifefish from South America, and the Mormyridae or elephant fish from Africa (Figure 16). As Darwin suggested, there is striking convergence in the evolution and use of electric signals in these two groups of fish. It has been calculated that two groups have origins somewhere in the region of 100–120 million years ago, which rather tantalizingly is around about the same time that Africa and South America split apart as the southern Atlantic Ocean formed.[6] Initially, a passive electric sense existed or arose in these fish, but then over a period of between 16 and 26 million years full electric organs for active electro-production evolved in both groups.

Lissmann also showed that species of electric fish can be broadly divided into two according to the type of EOD produced. The first, referred to as wave-type fish, produce EODs that are more or less continuous in time, with rises and falls in electric signals (voltage) in the form of waves of activity. By contrast, the second, pulse-type fish, produce more discrete 'pulses' of EODs separated in time. The form of the EODs often consists of a specific number of phases. Some fish produce EODs of just one phase, whereas others produce bi-phasic signals with a rising phase followed by a trough. Other fish produce EODs with multiple phases. The specific phase components can provide important information in communication. Electric discharges

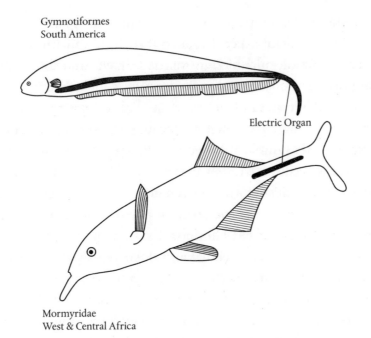

Gymnotiformes
South America

Electric Organ

Mormyridae
West & Central Africa

Figure 16. Electric knifefish from South America (above) and elephant fish from Africa (below), with the position of their electric organs indicated.

are also very flexible within individuals. When in the presence of other fish that are producing similar EOD signals, individuals will often change their own EOD properties to reduce overlap. It seems electric fish can change aspects of their EODs to avoid interference from others, enabling their sense to work more effectively. We often call this a jamming avoidance response, and it is not dissimilar to the way in which many bats modify their echolocation calls to avoid overlap with the cries of others, as well as noise in the environment (such as high-frequency insect calls).

The similarities between the types of EODs (e.g. wave and pulse forms) made by the knife and elephant fish often reflect convergent evolution due to similar types of habitat and social interactions. Curiously, such convergence in signal form also involves considerable convergence at the molecular level. With advances in molecular biology, scientists have been able to show that non-electric fish a gene called *Scn4aa* is expressed in skeletal muscle.[7] This encodes features of special sodium channel proteins that facilitate the movement of ions across cell membranes. In electric fish, however, this gene is no longer expressed in skeletal muscle. Instead, it is now expressed in the cells of the electric organs, in a modified form. The remarkable thing is that the same mutations to this gene, and corresponding amino acid substitutions, have been selected, altering the same regions of the sodium channel proteins, in both knife and elephant fish independently. This shows that, just as with the echolocation abilities of bats and dolphins, and the infrared sense of snakes, convergent evolution can occur not only at the level of observable features, or phenotypic level, but also at the molecular level. During that period of 16–26 million years, changes in physiology based on similar alterations in their ancestral genes allowed the evolution of an electric organ in the two main groups of electric fish.

As well as producing electricity, electric fish are highly sensitive to weak electric signals (some fish can detect signals of less than 10 mV/cm) and have been found to have sophisticated sensory apparatus for picking up EODs. In many ways this discovery was not surprising. Producing electricity is only one part of the task for an active electric sense; the fish needs to be able to detect and interpret the information produced, so it must have evolved dedicated electroreceptors for measuring its own EODs and those of other fish. But prior to the 1950s, when Lissmann and others began studying the EOD properties of electric fish, it was largely uncertain how they did this. Evidence from nerve cell recordings shows that the sensory receptors involved are small pits in the skin, with a group of sensory nerve cells at the base, each acting to measure changes in voltage across the skin. Like sharks and rays, electric fish can respond to passive electric information in the environment, such as from the body and muscle actions of prey. To do this, they have one type of sensory cell called ampullary electroreceptors, which detect weak electric fields (around 0.01–1 mV/cm) at low frequency (1–100 Hz). However, they also have a second type of sensory cell called tuberous electroreceptors, which specifically detect features of EODs. These are sensitive to stronger electrical information of higher frequency (0.1–10 mV/cm between around 100–5,000 Hz). The tuberous cells can be further subdivided based on their role in detecting information used in specific tasks. For example, some cells are used in object detection (essentially analysing changes in self-generated EOD signals), whereas other cells are used in communication (detecting the EOD signals of other fish). That is, receptors are modified for picking up types of information that underpin different tasks.

We are used to thinking about how visual and auditory signals are used for tasks such as mate choice and territoriality in many animals. The gaudiness of the peacock's tail in attracting a peahen and the call of a great tit in defending a territory are two classic examples. Our focus on these is partly because we can see and hear such signals ourselves. The sensory world of other animals is very different, but it turns out that electric fish exploit electricity in a not entirely dissimilar way to peacocks. The electric sense plays a major role in communication between electric fish and it has had a major influence on their evolution.

In 1972, Carl Hopkins, then at Rockefeller University, showed that a species of fish, *Sternopygus macrurus*, that lives in rivers and swamps in places like Guyana in South America, has an electric discharge that is not only different to those of other species in the area, but that its EOD features also differ between males and females.[8] Males have slower electric discharge patterns (an average frequency of 67 Hz), whereas those of females are of higher frequency (120 Hz). Males in particular also use different signals during courtship. Experiments playing back electric information of different frequencies to a small number of potential mates showed that these differences mattered in communication; animals respond most to playbacks resembling signals of their own species.

Further work has shed light on how electric signals are used in spawning behaviour.[9] In some species of South American fish, individuals living in small groups form dominance hierarchies and have a variety of other social interactions, and these are correlated with features of the electric signals used. Generally, dominant males tend to be the largest individuals with the lowest frequency EODs. By contrast, dominant females tend to be the most territorial and have higher frequency EODs than other females. Males also produce modified EODs in the form of more rapid 'chirps', and these play a role in aggression and in dominance between males. They also make calls towards females to encourage spawning behaviour, especially in the evening, when males chirp 'incessantly'. A male may need to chirp for at least an hour before a female will consider spawning with him.

The EODs of electric fish play a crucial role in mating, conveying a wide range of information through variation in features such as the timing, amplitude, and phase. The pulse-type knife-fish makes a good example. Many of these fish have 'bi-phasic' EODs, consisting of a peak and a trough (Figure 17). The form of the first, rising phase can be specific to the species, helping individuals to associate and mate with the 'right' species. The second phase on the other hand is often specific to the sex of the signalling fish. In other words, many electric fish are sexually dimorphic, with the males having a different EOD structure to the females.

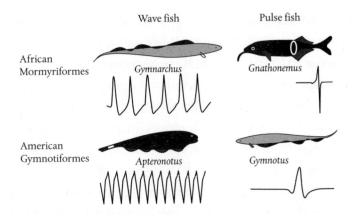

Figure 17. Electric fish broadly come in two main types, based on the general features of their electric organ discharges (EODs). Wave-type fish (left) show relatively continuous patterns of EODs over time, whereas pulse fish (right) show more discrete EODs separated in time. Each EOD can be characterized by features such as its amplitude or intensity, how many phases it has (peaks and troughs), and how long each phase lasts.

This is analogous to the difference in plumage colours between males and females in many birds.

One particular knifefish, the pintail knifefish (*Brachyhypopomus pinnicaudatus*), has been studied in detail. This is a relatively dull brown species that occupies the slower moving waters at the edges of rivers and streams. During mating, male fish change their signals depending on the context. When courting a female, they produce bursts of accelerating signals to induce females to mate. They then switch to the 'chirps' described above at the spawning stage. As with many other species, the EODs between males and females are dimorphic, with males having a more exaggerated signal. Scientists have shown that females choose males with which to mate based on features of the male EODs.[10] When given a choice of male, detectable only by the males' EODs, females associate more with males producing EODs of higher amplitude (that is, stronger) and of longer duration. Both these features correlate with male body size. Quite why females prefer bigger males is not fully known, but they certainly do seem to select males based on their EOD properties, just as other species choose mates based on the intensity of colour in visual signals or frequency and complexity of songs.

Perhaps the basis of female choice in pintail knifefish is that the strength and duration of a male's electric signal indicates his quality as a mate, that is, his biological fitness. Male EODs are energetically very expensive, consuming on average 22 per cent of the daily energy needs of the fish (compared to just 3 per cent of the energy budget of females), and the more intense the EOD, the greater the cost to the male.[11] So it is possible that only males of high quality, or those in very good condition, can produce strongly exaggerated EODs. Naturally, this may also be connected to male body size. More exaggerated EODs may also carry a cost in potentially attracting predatory species that can eavesdrop on the male's mating displays, much like the plumage of peacocks and birds-of-paradise risks attracting predators and can impede effective flight.

EODs do not just facilitate mate selection within a species, but also play a key role in choosing to mate with the right species. Given this, it is little surprise that EODs have profoundly shaped the evolution of these animals as a whole. Some parts of Africa have around 200 species of electric elephant fish. This is a lot when considered in comparison to approximately just ten species of their closest non-electric relatives. The electric sense has driven large evolutionary radiations of these fish, with lots of new species forming in the same area, sometimes in just a couple of million years (a relatively short space of time in evolutionary terms). In evolution, there is often thought to be strong selection to avoid mating with the wrong species, and here, species living

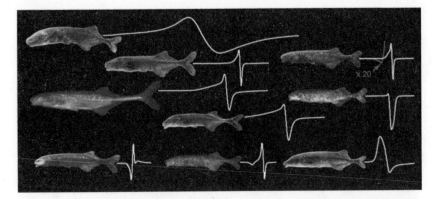

Figure 18. Diversity in the EODs from fish from Equatorial Guinea and Gabon, belonging to closely related species living in the same area. These fish have undergone radiations in species abundance in the same locations, associated with changes in the structure and duration of their electric signals.

in the same area have EODs that are more different in structure than species that live in different areas (Figure 18).

There are some animals that at first seem too bizarre to be real. The platypus (*Ornithorhynchus anatinus*) is one such (Figure 19). In fact, it is so strange that early evidence for its existence in the form of skeletons and pelts were often considered a hoax—the cobbled together parts of a beaver-like animal and a duck. If its strange appearance were not enough, platypus have ankle spurs that in males are associated with the production of venom from specific glands. The venom is quite potent—apparently strong enough to kill a medium-sized mammal like a dog—and is probably used in defence (and perhaps in male competition). And the platypus is also unusual in being an egg-laying mammal—a monotreme, along with species of echidna. Aside from having these curious traits, its remarkable bill attracts immediate attention. To find food, the platypus dives underwater and rummages in the riverbed

Figure 19. The platypus (*Ornithorhynchus anatinus*), showing its famous bill, which is packed with receptors for detecting mechanical information (such as water movement), and electroreceptors for finding hidden prey.

for prey. Its vision is not very good (in fact it often keeps its eyes closed when foraging), and the water is often murky, yet its bill is a highly specialized sensory organ. As the platypus swims, it sweeps its bill from side to side, as if scanning the environment around it. For good reason: on the bill structure are thousands of cells that respond to touch, pressure, and to electricity.

Since the recognition of the electric sense in various fish, zoologists had wondered whether such a sense was to be found in any other animal groups. In 1986, scientists demonstrated that the platypus had an electric sense too.[12] They showed that platypus making characteristic sweeps of their bill close to the river bottom were capable of pinpointing the position of a miniature 1.5V battery. In fact, in their experiments, the platypus had a clear preference for a live battery over a dead one—or even a piece of shrimp tail. Even more impressive was the fact that the animals could detour around a plastic plate suspended in the water when the plate was surrounded by an electric field, whereas they routinely bumped into the plate when the field was turned off. So, evidently, the platypus can use their sense both to avoid objects in the water and to locate small prey items. In the case of the shrimp, the tail flicks made when they tried to escape produced sufficient electrical activity to be detected by the platypus. By recording activity in the brains of individual platypus while stimulating parts of the body, researchers showed that their electric sense was located on the bill. They postulated that the receptors were likely to be linked to long known gland ducts found on the bill, and also found in the echidna. The presence of electroreceptors there was confirmed soon after.

The shape of the bill is also part of the adaptation. Its wide, flat form enables the large number of receptors to be packed in and across its surface, and for the bill to be swept across and make contact with parts of the substrate as the animal forages. The bill also reflects another change to enhance the senses of the platypus: the animal lacks functional teeth, unlike a similar but extinct species that has been recorded. It seems the loss of teeth allowed the bill to accommodate greater numbers of electrical and mechanical receptors and their associated nerve cells. The roots of teeth would have taken up valuable space, which is better invested in the electric sense.

Not long after this work on the platypus, scientists showed that the echidna (Tachyglossus aculeatus) also has receptors on the tip of its snout that respond to electric information.[13] The animals could be trained to discriminate between water filled containers based on the presence or absence of an electric field. At the time of this discovery, the echidna was the only known terrestrial animal that had been shown to use an electric sense. The researchers postulated that their digging behaviour for insects and other food could mean that they encounter muddy, moist substrates where an electric sense can work. For most terrestrial animals, however, the sense is not a viable option.

Even if the echidna does have an electric sense, it seems to be quite limited. On the platypus bill are around 40,000 receptors that respond to the tiny amounts of electricity given off by its prey. In comparison, the echidna seems to have a mere hundred of these receptors, and presumably a much more rudimentary power to process electric information. Interestingly, another living monotreme, the long-nosed echidna (*Zaglossus bruijnii*), probably also has an electric sense, and potentially up to 3,000 receptors, which it may use to locate worms and insects on the forest floor. Unfortunately, the species is rare and elusive, found in the mountains of New Guinea, and therefore not well studied.

The monotremes are a remarkable group of animals—we might mistakenly think of them as a sort of transition between reptiles and modern mammals, in that they share reptilian characteristics like egg-laying and reptilian limb patterns, but mammalian traits such as suckling young with milk. However, to think of them as 'primitive' or transitional would be wrong. Since their ancestors split from those of other mammal groups (over 100 million years ago) they have evolved in their own specialist way, including traits such as an electric sense that, as far as we know, are missing in other mammalian groups. The platypus in particular has in its bill a sensory organ that is highly advanced and one that few other mammals could compete with.

The echidna tells us that the electric sense need not be quite so restricted to fully aquatic animals as we might first assume. However, the better conductivity of water must surely restrict how widespread an electric sense can be in nature, and there seems little doubt that it reaches its pinnacle in electric fish.

Alongside those fish, the platypus, and a few other aquatic animals, it may be more commonplace in invertebrates (including crayfish). Yet this overall perspective is based largely on a more traditional consideration of what an electric sense is and how it should work. Recently, we've had to re-evaluate our thinking as evidence has come to light of animals using electric information in a completely different manner, based on sensors used traditionally in other modalities.

In 2013, Dominic Clarke and others at Bristol University discovered that bumblebees are able to detect electric fields associated with flowers.[14] The bees may even use this to guide which flowers they should visit. When a bee flies, it gradually accumulates a positive charge, from static electricity. In contrast to bees, flowers tend to have a negative charge. The ground itself accumulates a negative charge in response to the positive potential of air (related to the so-called 'global atmospheric electric circuit'), and objects connected to the ground, including flowers and trees, as a result accumulate negative charge too. The upshot is that when a bee visits a flower the differences in charge can cause pollen to be 'attracted' to the bee. Moreover, when a flower has recently been visited, the electric field changes, meaning that a new visitor may quickly be able to know if that flower has been visited already. The bees can detect movement and distortions produced by changes in electric charge when they visit the flowers, and seem to tell even the shape of flowers based on the geometry of the electric field.

Bees have been shown to sense electric information by using hairs that respond to mechanical cues. On interacting with an electric field, the hairs are physically distorted and moved, leading to nerve impulses that the bees are seemingly able to detect. Therefore, bees are able to detect electric information even in the complete absence of any conductive medium for the electricity, such as water, by obtaining it indirectly from mechanical information. And other recent work indicates that honeybees may be able to detect aspects of the 'waggle dances' (made by workers to tell nest mates the position of food sources) based on electric information as the dancing bee moves its body around. What we do not yet know is whether this sort of electric sense is found widely in other animals, though when there is one example, there are often many.

We can try to imagine the extension of a sense we already have, such as ultraviolet vision, because although we cannot see UV light we at least know what vision and colour perception feel like and how we use our visual system. But when we lack a sense entirely, as in the case of the electric sense, things take on a new level of difficulty. In spite of this, great progress has been made in understanding not just how the electric sense works in many animals, but the wonderful variety of ways in which it is used, the complexity of information acquired with it, and even the key role that an electric sense has played in driving the evolution of species diversity. If nothing else, the electric sense shows us that even in the sensory worlds of animals that are beyond our personal reach, the refinement and sophistication of those senses can be in line with any sense we might be more familiar with. In Chapter 5 we return to a sense that we do have, a sense of touch, but find that our ability here barely comes close to that of many other species.

CHAPTER

5

STARS OF THE TACTILE WORLD

For the harbour seal (*Phoca vitulina*), capturing moving prey in turbid and often murky waters is a challenge requiring great skill and dexterity. It also demands exquisite sensory apparatus to locate and pinpoint targets like fish. The water currents made by the fish as they swim enables the seal to track them, but much more than this, the predator can determine the size, movement, direction, and even shape of the prey it's hunting.[1] The ability to sense the patterns of water movement as a fish swims past, which persist momentarily after it has gone, is conferred by the seal's whiskers (called vibrissae). These can detect small changes in pressure and touch. By combining information from multiple whiskers, and the sensory cells at their base, the seal obtains a great deal of information about the object it is tracking. Furthermore, large areas of the brain are devoted to processing mechanical information and guiding the seal's behaviour, driving its wonderful ability to track and grasp its meal.

The risk of being attacked and eaten is clearly a serious problem for most animals, and they go to great measures to avoid this. Some animals use camouflage to merge with their background, others take safety in numbers, or even have defensive armour and poisons. But a simpler and common tactic is to hide in mud or sand, or behind objects, in order to be overlooked entirely. Unfortunately, this is often not enough. We know that sharks can detect a buried flatfish by virtue of the tiny amounts of electric information the fish's body produces. Not all animals can find prey this way, but they have other remarkable routes to success. Incredibly, the harbour seal can find a buried flatfish by nothing more than the water flow created by the fish breathing. Even with the fish hidden on the seabed, the seal's whiskers are so sensitive to water movement that they can detect the intermittent and small amounts of water flowing out of the flatfish's gills as it breathes. After knowing this, we might rightly pause in awe of the sensory abilities animals use to find their food, but we can perhaps also spare a thought for the poor flatfish in trying to stay alive in the face of sharks and seals. In this chapter, we explore the supreme level of refinement found in many animals for analysing tactile and pressure information, starting with a species that has also revealed much about how the brain processes information.

In the wetlands of North America lives a small mammal, weighing just 40–50 g, which spends much of its life underground in burrows. Occasionally, it comes to the surface to forage for food in mud and water, often in and around streams, as well as searching in the drier soil of its tunnels. It is an adept hunter of invertebrate prey, yet its eyes are small and there is very little nervous system devoted to vision, so its sight is probably rudimentary at best. And although its hearing may be better than in some similar species, this also does not seem to be especially valuable. Speculation on the use of other more unusual senses, such as potentially an electric sense, also seem unfounded. The most obvious feature of this animal is the structure that gives it its name, its nose. It is the star-nosed mole (*Condylura cristata*), and its sensory organ is a marvel of sophistication (Figure 20).

The mole's star-shaped nasal organ is just 1 cm in diameter, smaller than the width of an adult thumb, with twenty-two fleshy and mobile appendages that fan out around the nostrils. The nose structure was first noted in the late 1800s, and by the time William Hamilton described the natural history of the star-nosed mole in 1931 it was established that the nose is not primarily used for smell, but instead for touch.[2] Details of the star's function have only really been investigated since the 1990s, albeit now studied in depth. In the past three decades, scientists have uncovered how this sensory organ works, and how it enables the mole to live and find food in its environment. Alongside a range of colleagues, nobody has done more to uncover the secrets of how the nose works than Kenneth Catania, formerly of the University of California, San Diego, and now at Vanderbilt

Figure 20. The star-nosed mole (*C. cristata*) showing its eponymous nose structure.

University in the USA. Studies of highly specialized sensory apparatus often attract neurobiologists because these systems reveal much about brain function and organization, and how animals process information to guide behaviour. Like the barn owls, and hearing in general, the star-nosed mole has played an important part in our understanding of sensory biology.

Although the appendages on the nose might look similar, they are not analogous to fingers—they are not used to manipulate items and contain no muscle. Instead, they are used purely for collecting tactile information. When searching for food, the nose is constantly moved about, touching the tunnel area, and hitting or missing food items. The appendages are quickly swept backwards and forwards and side to side, interacting with different parts of the substrate at astonishing speed: up to thirteen times a second.[3] If they touch a prey item, such as an insect or worm, the prey is quickly consumed. The twenty-two appendages that comprise the nose organ span out from around the nostrils, with eighteen of these, nine on each side of the nose, being longer and forming a ring that usually first detects the food. Then, one of the middle appendages (number 11), which is directly adjacent to the mouth, is used to repeatedly touch the food before ingestion, to allow detailed assessment of the prey item. In other words, some of the appendages are used for initial detection, while others are for identification. The whole process from detection to ingestion takes just half a second.

Such incredible speed of prey detection and consumption is achieved via considerable processing power. Every appendage is covered with a honeycomb of special structures called Eimer's organs, each being a raised bump on the skin with five to ten underlying sensory nerve endings.[4] These structures are found in other mole species too, but in the star-nosed mole they are smaller, more organized, under thinner skin, and occur in higher number. All in all, about 25,000–30,000 Eimer's organs cover the twenty-two appendages, around 1,000–2,000 per appendage, albeit with fewer on the central ones since they are smaller (Figure 21). Despite being just 1 cm across, the nose has five times more sensory nerve fibres (100,000) than exist in equivalent cells for detecting touch on an entire human hand (around 17,000). Each Eimer's organ is only 15–16 μm in diameter, which is equivalent to about three to four red blood cells lined up. These features give a stunning level of precision for detecting and analysing objects in great spatial detail.

The sensory cells in the Eimer's organs respond to sustained pressure of touch, and to sensations of touch that are on and off. Given the relatively small size of most objects that are of relevance to the mole (e.g. invertebrate prey), the patterns of touch that arise in a single appendage will produce strong stimulation from multiple receptors when an object is encountered. In contrast, the other appendages will have little stimulation, until the mole

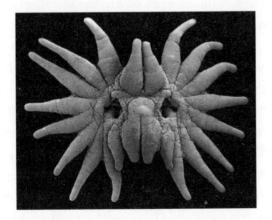

Figure 21. The remarkable nose structure of the star-nosed mole (C. *cristata*), showing the eleven appendages on each side of the nose in rings around the nostrils. The bumps are domes comprising each of thousands of Eimer's organs distributed across the structure, which are highly sensitive to pressure information.

moves its nose and another appendage touches the object instead. In this way, the mole can gain information about the location of the objects around it based on which part of the nose is stimulated. The mole can then also deduce fine-scale information about an object it has detected. The nerve cells respond to intricate detail, including gathering information about direction and movement patterns of stimulation, especially those coinciding with the speed at which the nose would normally be moved by the mole as it shuffles around. Owing to the number of Eimer's organs, their tiny size, and the way that the sensory cells respond to patterns of stimulation across parts of each individual Eimer's organ, the mole obtains exquisite detail on texture, almost to a microscopic level.[4] Recent work has also shown that the arrangement of Eimer's organs, with regards to their density, is somewhat analogous to the human hand: close to the mole's nostrils there are relatively fewer organs, but as you move along the appendages to the tips, the number increases. This is similar to the way in which the number of touch cells increases between the palm of our hand and our fingertips, where we are most sensitive.

Gathering such refined levels of information is only part of the story, however, because the mole's brain must still somehow process and make use of this exquisite detail. The mole needs to have a serious level of brain power allocated to deal with it all. Catania and other researchers in this field work out how the brain processes information by making recordings of nerve cell activity in the brains of anaesthetized animals, while gently stimulating parts of the body, including areas of the skin on the nose. Sometimes particular areas of the brain are found to become active when a specific region of the sensory organ is stimulated. In fact, almost half of the entire brain of the mole turns out to be dedicated to processing sensory information from the nose. The level of devotion to touch in the star-nosed brain is so great that even parts of the brain normally used for vision have been taken over by the need to process tactile information.[5] More than this, the brain is also anatomically arranged in a very particular way to facilitate analysing what the mole's nose has detected.

As with many other animals, there are two hemispheres of the brain, with each processing information coming from the opposite side of the body. A large area in the cerebral cortex (a somatosensory area) processes touch information. Within this region are morphologically distinct patterns of subdivisions, referred to as modules, each of which is directly related to one of the nose appendages.[6] In this layout the area related to appendage 1 are located next to an area for processing information from appendage 2, and so on. Thus, we end up with eleven distinct 'stripes' of tissue in the brain in each hemisphere, each corresponding to one of the different appendages, and arranged rather like a pinwheel. Different areas are then connected with one another and to other brain regions for further analysis (Figure 22).

Information from the nose is relayed and presented to the relevant brain areas in a point-by-point mapping, so there is a neural map of the sensory receptor surface of the nose in the mole's brain. This is very much like the way in which the barn owl creates a spatial map of sounds that it detects around its body in its brain. By having such an arrangement, each appendage can act as an independent module, and when one region fires in the brain it tells the mole which appendage is stimulated, and hence where the food item is (as well as abundant information regarding its shape, size, texture, and so on). The mole is able to consume prey items so rapidly because the brain and sensory system has so much power to process the relevant information. In fact, it's hard to take in just how impressive this is, and how rapidly

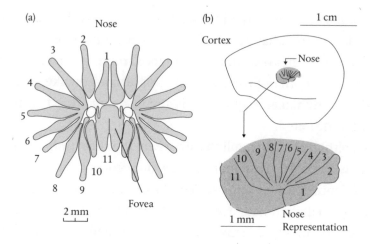

Figure 22. The arrangement of the eleven appendages on each side of the star-nosed mole's nose map to dedicated areas or 'stripes' in the brain, each of which is devoted to processing information from each appendage.

impulses from the sensory receptors travel to somatosensory areas of the brain, and then back to the motor systems controlling the mole's behaviour. Neurons in the somatosensory cortex respond to touch occurring on the star within just 12 ms, and the mole can locate, identify, and eat a prey item in as little as 120 ms. By comparison, it takes humans 100 ms to blink.[7]

Just like a variety of sensory organs in other species, the Eimer's organs are associated with receptive fields in the brain. Touching one Eimer's organ directly stimulates a receptive field, but if another Eimer's organ close by is touched then the receptive field of the neighbouring organ is inhibited. In this way, the receptive fields respond to contrast information and relative patterns of stimulation, and are more activated when one region is stimulated but another region close by is not. Animal visual systems often have receptive fields that do something analogous in responding to contrasting arrangements of light and dark in the environment that fall on the eye's receptors. The mole's receptive fields can sometimes equate to areas of skin averaging less than 1 mm². Much akin to similar processes in vision, and in the barn owl's hearing and spatial maps, such precision enables the brain to work out the edges and boundaries of objects, judge contrast, and encode where things are occurring in space to a fine resolution.

There are also direct links between the way that the mole behaves in finding and analysing food, and how the brain is set up to process touch information. Once a mole makes contact with a food item, through any of the nose appendages, it tends to then preferentially use the eleventh appendage to explore the food item further. The eleventh appendage is small and has fewer receptors on its surface than the other longer appendages, but it actually has a larger area of brain and more nerve cells devoted to processing its responses. High-speed video recordings of moles foraging on Plexiglas showed that during foraging, the mole first searches at random, sweeping the nose about and using all the appendages to detect objects, especially the longer ones since they have more sensors. Once an object is detected, the mole shifts focus and uses the smaller eleventh appendage to explore it, and to utilize the correspondingly greater brain power to work out fine detail regarding texture and shape for identification and other purposes.

There is one other aspect to the way that the star-nose mole processes touch information that helps us to understand how animal brains more broadly encode features of the world around them that are of special importance. The nose itself makes up a small fraction of the overall skin surface of the mole, yet the somatosensory cortex in the brain is dominated by sensory input from the nose. Likewise, the forelimb of the mole is heavily overrepresented in the brain, again because it is important in burrowing, detecting vibrations, and other similar things. Over representing parts or regions of the body with regards to how the brain processes sensory information is a phenomenon called 'cortical magnification' (Figure 23). It is something that happens in the human brain too. For instance, our hands and lips have a greater brain representation for touch than does our entire back, due to their importance. The relative representation relates to the increased number of nerve fibres that innervate the surface of the sensory organ; that is, the area in the brain allocated is proportional to the density or number of receptors found in the sensory structure itself. Many similar examples have been found in other species too, where different senses are of special importance to their ecology and way of life. Returning to echolocation in bats, sound frequencies of special significance have an abundance of

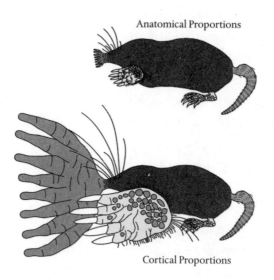

Anatomical Proportions

Cortical Proportions

Figure 23. On the top is an illustration of the star-nosed mole showing its body accurately with regards to respective anatomical proportions. Below is an illustration of the body with different regions made smaller or larger to reflect cortical magnification in the brain, whereby there are much larger areas of the brain devoted to processing touch information from the star and forelimbs than from other body areas.

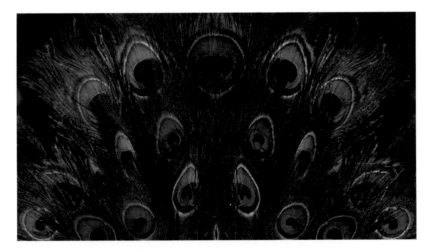

Plate 1. A simulation of how avian vision may allow the colours of some objects with ultraviolet reflectance, such as the feathers of a peacock to be more vibrant and colourful than would be seen by human eyes (left).

Plate 2. The ornate jumping spider (*Cosmophasis umbratica*), showing the iridescent colours found on the male, many of which reflect ultraviolet light.

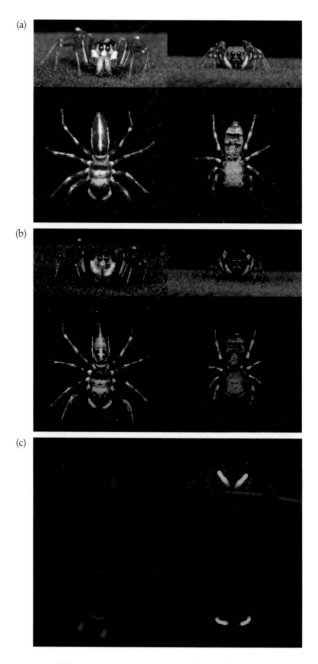

Plate 3. Images of the ornate jumping spider under (a) human visible, (b) UV light, and (c) UV-induced fluorescence. Images show males on the left of each pair and females on the right. In mate choice experiments, males prefer females with fluorescence whereas females prefer males with UV markings.

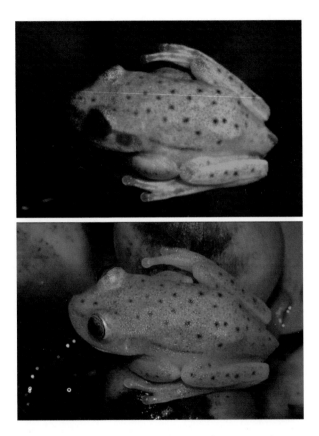

Plate 4. The South American polka dot tree frog (*Hypsiboas punctatus*) is a yellow-green colour with small spots under normal lighting, but under ultraviolet light it glows green with fluorescence. Whether this has any function is unknown, but it may make the frogs more visible to other individuals under twilight conditions.

Plate 5. A typical deciduous woodland scene looks very different to animals with different types of colour vision. To an animal with no colour vision (e.g. some nocturnal primates) the world appears as shades of grey (left images), whereas to a dichromatic animal such as a dog, there is colour but it is restricted to what humans would see as yellows and blues (centre images). To a trichromatic human, differences between greens, reds, and browns become apparent (right images). The images below show the same effect with a yellow lemon and red pepper set against green grass.

Plate 6. Mantis shrimp are not only exceptionally colourful but also have remarkable eyes. These can include at least 12 receptors sensitive to different wavebands of light, and the ability to see polarized light. Many of the receptors are housed in a special area of the eye, called the midband, which can be seen as lines running horizontally across the middle of eye.

Plate 7. The golden silk spider (*Nephila clavipes*) measures vibrations travelling across its web with sensors on each of its eight legs to detect where objects are in the web and what they may be.

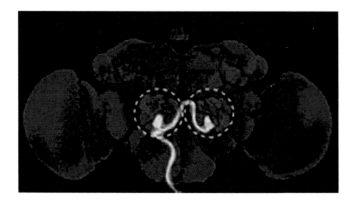

Plate 8. Processing of odour information in the fruit fly. Sensory information corresponding to different odour molecules is relayed via sensory neurons (yellow and green lines) to the antennal lobe area of the brain. Here, different outputs from different odour receptor types are compared and the pattern of activity used to encode the specific odour present. Information is then transferred (purple lines) to higher processing centres, including those that control behaviours such as flight and learning.

Plate 9. The appearance of a fence from Cheltenham Racecourse, UK, to human (top) and horse (bottom vision). The orange visibility markers are clear orange against the green to a person (e.g. jockey) with normal colour vision, but to a horse the entire scene appears as shades of green-yellow.

Natural   Loud natural   Ship

Plate 10. The shore crab (*Carcinus maenas*) is able to change its brightness to better match the background for camouflage, for example, becoming lighter when on a white background. Experiments have shown that shore crabs are adversely affected by ship noise, and that they do not change colour properly when exposed to this. This image shows three crabs that were initially dark and then became lighter when kept on a white background. However, crabs exposed to ship noise (right) changed colour less than crabs that were exposed to natural seashore sounds at a normal level (left) and at a higher intensity (centre).

Plate 11. Green LED lights added into fishing nets have been shown to reduce the chance of capturing turtles while maintaining capture of target fish species.

hair cells for detecting them in the ear, and these are in turn allocated considerable brain area and processing power.

While the above is largely true in the star-nosed mole, there are some deviations from this general trend. Unlike in certain other species, the area of mole cortex for each appendix is not exactly proportional to the number of receptors on the appendages. Instead, the importance of each appendage in feeding behaviour also plays a part in allocating brain space. As we know, the eleventh appendage is overrepresented, despite having fewer receptors. This suggests that structures that are especially important in analysing prey are allocated more cortical space. Catania has suggested that the two eleventh appendages of the star are tactile equivalents to the fovea of many animal visual systems. In the fovea of our eyes there is a higher density on the retina given to cells used in things like colour vision and recording fine detail. Correspondingly, the receptive fields of the Eimer's organs encode smaller areas of touch space on the eleventh appendage compared to those on the other appendages, giving more fine-scale information. While there isn't a higher density of Eimer's organs on the eleventh appendage, there are more nerve cells feeding into the Eimer's organs than on the other appendages, allowing for more discriminatory power. So, there appear to be two factors in how much brain area is given to specific sensory organs or regions of them: how many sensory structures there are and how important that area or feature is in behaviour, and these do not always exactly align. Nonetheless, regardless of which factor dominates, the key thing is that animals devote more brain area to sensory systems, and aspects of them that are more critical in their lives for performing specific tasks. As such, the maps of the senses formed in the brain do not correspond to the size of the body parts but to the amount of sensory information received from them.

In other species of mole, touch information is also clearly apparent in large dedicated brain areas, but much less so than in the star-nosed form. Across different mole species, considerable area of the brain is frequently set aside for processing touch information from the forepaws, which suggests that sensory information from touch is processed and used for controlling digging behaviour. The star-nosed mole also has much thinner skin, and less well protected, more delicate Eimer's organs, than other species of mole that live in drier soil. The wet environment in which the star-nosed mole lives may have provided less abrasive conditions, freeing up the sensory system to evolve to be more refined. The species has specialized in eating invertebrates and is very active, so it needs to eat often, and its specialized organ allows it to find food rapidly in an environment that would otherwise be hard to exploit. Sensory systems evolve to work best in an animal's predominant environment, and enable it to exploit habitats and ways of life with which other species would struggle.

Rodents have also been widely studied to understand how sensory systems acquire and process touch information. They represent valuable model species to investigate the neurobiology of animal senses and how sensory processing guides behaviour. In mice and rats, for which touch is crucially important for survival, tactile sensory information is gathered not from the nose but from the whiskers or vibrissae. The arrangement observed in the brain of the star-nosed mole, in which sensory information from each appendage is handled in a dedicated brain area, is mirrored in the processing of touch information from the whiskers of rodents. Each whisker acts as a discrete sensory unit with its own allocated area of brain in the somatosensory cortex, and the number and density of sensory receptors found in mice relates directly to the degree of representation in the somatosensory cortex.

In other rodents too there is a well-demonstrated link between the importance of the main senses and the processing

Figure 24. The naked mole rat is a highly social species of mammal that spends most of its life underground. Studies of its sensory systems show that it has a highly developed sense of touch centred on its prominent teeth.

of that information. Naked mole rats (*Heterocephalus glaber*) are subterranean rodents that live in arid parts of eastern Africa. They are rather odd creatures, to put it mildly (Figure 24). Unusually for mammals, they cannot properly regulate their own body temperature, instead relying on moving to different burrows closer to or further away from the warm surface. They would not win any prizes for the most attractive animals in the world either, being wrinkly, pink, largely hairless, and with huge teeth and tiny eyes. Even more unusually for mammals, they are one of a very few vertebrates that show extreme levels of group living and sociality ('eusociality'). Fairly large groups of anything between 25 and 300 individuals co-exist, with a breeding female 'queen' and a set of workers that show divisions of labour and undertake different roles in the colony. They are, in other words, rather like mammalian versions of ants. This social aspect of their biology has been relatively well studied, but for our purpose here it is their sense of touch that is most interesting. Naked mole rats live underground and search for food such as plant tubers by digging networks of burrows. Their incisors are huge, and are used for digging as well as in behavioural interactions, given their highly social lives. The teeth are also remarkable sensory organs for analysing the environment.

Catania, together with Michael Remple, investigated the naked mole rat's sense of touch, and showed that many of the principles of the star-nosed mole's nose, and of rat and mice whiskers, also apply to the naked mole rat's front teeth too, perhaps even in a more extreme manner.[8] By making brain recordings and stimulating the teeth, they showed that one-third of the naked mole rat's primary somatosensory cortex is devoted to processing touch information from the incisors. Touch information arising from the front teeth is so important that areas of the brain normally used for other things have been taken over for processing this information too. Parts of the cortex that would be used for handling information from the facial hairs are now used for analysing the incisors. More dramatically, the somatosensory cortex has become greatly enlarged compared to that of lab rats, and now occupies most of the neocortex, which would normally be used for vision. This is loosely similar to the way in which some parts of the brain in animals can be 'reassigned' for other tasks when there has been the loss of one sense earlier in life. However, here, because evolution has had plenty of time to do this, the amount of brain area reassigned to tactile information from the incisors is considerable. Again, when we consider how naked mole rats live their lives, this all makes sense—they live underground in the dark, so retaining areas of the brain for a nearly useless sense (vision) when such neural power could be reallocated to processing valuable information (touch) would be a waste. However, as Catania and Remple have pointed out, exactly what sensory information is acquired from the teeth, and how it is analysed and used, is not yet clear, though it must be central to their livelihoods.

Almost any natural history documentary depicting the struggles for life on the African savannah usually can't help but mention the mass movements of migrating wildebeest. These large grazers travel great distances in immense herds to find grass to feed on. Along the way, they often have to cross wide murky rivers, packed full of lurking crocodiles. Cue scenes of wildebeest scrambling madly across the rivers and up the banks, while crocodiles snap and grab at them, trying to get a hold and pull unfortunate individuals down into the water. It might look like it, but the crocodiles are not simply relying on chance for an unwitting victim to stumble onto them. Somehow, they sense when one is close enough to strike.

There are twenty-three species of living crocodilians, including crocodiles and alligators. Occurring in a variety of semiaquatic habitats, ranging from rivers to even the ocean margins, they are adept at capturing prey underwater, and conditions where visibility is poor are of little hindrance. So how do they know when a suitable prey animal is close by, and how do they direct their strike towards it in good time and with high accuracy?

In 1996, a study by Kate Jackson and colleges from the University of Toronto explored the function of structures that are found on the jaws of crocodilians (in fact all over the bodies of crocodiles).[9] These features, called integumentary sense organs or ISOs, have often been used as traits to distinguish specific species, but while it had been suggested that they function as sensory organs, little was known about them. Some scientists had alternatively suggested that they were used to secrete substances, perhaps for waterproofing the body. Jackson and others studied the estuarine crocodile (*Crocodylus porosus*) and used micro-scopes to examine the fine detail of their ISOs. The organs are raised bumps in the otherwise thick skin of the face. Where the structures occur, they comprise areas of relatively thinner skin (about half as thick) stretched over the middle of the bumps, and immediately underneath the surface are various nerve fibre endings—a telltale sign of sensory organs. All in all, these and other features strongly suggested a function in the sense of touch, yet it was still speculative. Aside from secreting body substances, other suggestions for ISO function included that they were used as chemical receptors and perhaps for detecting water salinity. In fact, work early in 2000s initially showed some evidence for the latter, but more on that shortly.

A few years later, Daphne Soars from the University of Maryland showed clearly how alligators (*Alligator mississippiensis*) orientate rapidly towards droplets falling silently into the water in darkness, and that mechanical or pressure information was key to ISO function.[10] When single drops of water fell into tanks, the alligators would turn either their head or their whole body towards the source of the ripples, but they would only do this when their head was at the air-water interface, and not when they were submerged or entirely out of the water. This is significant, because the water ripples travel across the surface. In a simple but elegant manipulation, when Soars covered the ISOs on the head with plastic, individuals stopped responding entirely. Underneath the ISOs numerous nerve endings were again observed, linked to branches of the trigeminal nerve—a key nerve that is specialized for use in a range of important verte-brate senses, from electroreception in the platypus to infrared detection in snakes. Next, Soars created a range of surface waves and showed that the trigeminal nerve responded to the presence of water ripples, with stronger waves causing more stimulation.

Further understanding of the nature and function of ISOs in both crocodiles (the Nile crocodile; *Crocodylus niloticus*) and alligators (the American alligator; *A. mississippiensis*) came from work in 2012 by Duncan Leitch and Catania.[11] They used microscopy and made recordings from nerve cells to analyse the ISOs on both species, and estimated that there were around 4,000 organs in the alligator and more than 9,000 on the crocodile (Figure 25). These ISOs contained a range of nerve cells and mechanoreceptors. The ISOs were often associated with receptive fields and were highly sensitive to different patterns of changes in pressure. The receptive fields were often smallest, giving the finest detail, on the jaws and nearer to the teeth. By using high-speed cameras, sensitive under infrared light, the scientists could record the responses of the two species to water vibrations in total darkness. The animals were able to orientate towards the source of objects or water being dropped into the tanks, and they would frequently sweep their head around searching for the sources of the vibrations. When contact was made they would rapidly, within 200 ms, bite the object and determine whether it was food or not.

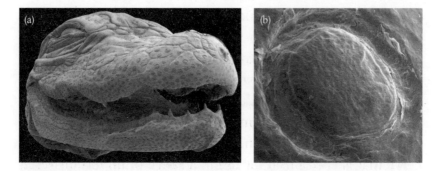

Figure 25. The head of an *A. mississippiensis* hatchling under a scanning electron microscope, showing the distribution of the integumentary sensory organs that comprise small bumps of thinned skin, with a single organ shown enlarged on the right, under which are sensory cells to measure mechanical information, such as from water movements.

A number of ISOs were found that responded rapidly to vibrations between 20–35 Hz, neatly fitting the pattern of water surface ripples that would be made by potential prey. The sensory organs are incredibly refined, and the nerves fire when the bump is compressed by as little as four-millionths of a metre. In fact, they are much more sensitive than primate fingertips, and recent work has shown that crocodilians are so good at detecting a range of surface waves in this frequency region that their aptitude is comparable to other hunters with acute vibration senses, such as fishing spiders.

The heavily built bodies of crocodiles and alligators belie a high sensitivity, being able to detect the slightest changes in touch and pressure. Leitch and Catania's work illustrates that the crocodilian ISOs are crucially important to their way of hunting, not just detecting water ripples of their prey and discriminating those from other vibration sources, but also knowing when and where to strike, and even being able to manipulate captured prey in the mouth. In alligators, most of the ISOs are found around the teeth, mouth, and jaws, which suggests that they use their sense of touch for manipulating and identifying prey. Yet crocodilians are also known for aiding their young when hatching, by gently helping to crack open the eggshells and carrying their young around. The sight of a huge mother crocodile pick up a baby in her mouth to carry it to water often looks like it will end in disaster, but these animals can have a light touch too. Given their enormous strength, they must have an impressive sense of touch to do this so gently.

A few years prior to Leitch and Catania's work, scientists had found evidence that the morphology of crocodile ISOs seemed to change in response to water salinity, suggesting that the function of ISOs may not be entirely straightforward. This would make sense in crocodiles because they often inhabit both fresh and salt water, and so must sense the environment they are in. As it happens, we know from recent discoveries that the ISOs are in fact multi-sensory organs, able to respond to changes in temperature and pH as well as mechanical changes. Scientists studying the Nile crocodile and the caiman (*Caiman crocodilus*) have undertaken detailed investigations into the structure and physiology of the ISO organs, as well as making further recordings from the sensory nerve cells when they are presented with different stimuli.[12] The ISOs of both species have receptors that responded to changes in pH from acid to alkali, to mechanical information via touch, and to thermal information corresponding to a range of temperatures. This sensory apparatus is associated with the expression ('activity') of genes that were already known to code for specific sensory channels for these functions at the molecular level. However, curiously, the ISOs did not respond to changes in salinity. Indeed, Leitch and Catania did not find any responses of the ISOs to salinity either. Overall, the ISOs have several sensory functions, but perhaps not in how they detect water salt content—that ability likely comes down to some other sensory system.

Interestingly, the morphology of ISOs leaves telltale features (hexagonal patterns and texture) in skulls, and these can be used to identify the presence of similar sensory systems in other species, including fossil forms. When Soares and colleagues analysed skulls from living and extinct reptiles, they found that present-day lizards do not have this type of sensory system. Looking at fossils, neither do extinct species of crocodilian that are thought to have been fully terrestrial. By contrast, extinct semi-aquatic crocodilians do show evidence for an ISO system, and this seems to have arisen sometime during the Early Jurassic, some 180 million years ago, and been maintained in crocodilians ever since. More unexpectedly, a recent discovery suggests that

ISOs may have even been present in one rather iconic group of extinct animals: tyrannosaurus dinosaurs. A study of a late tyrannosaur fossil (*Daspletosaurus horneri*), 75 million years old, from the Cretaceous appears to show bone textural features and structures that are similar to the ISO organs of crocodilians.[13] While the connection remains speculative, the implications are fascinating because this discovery implies that, despite being ter-restrial, tyrannosaurid dinosaurs relied considerably on infor-mation from touch. Rather than using the facial organs to detect water movement, they may have responded to direct touch and rubbing, perhaps when interacting with objects and with one another, even using their facial regions in mating interactions, picking up eggs, and in detecting aspects of their nest and local environment, including temperature. It is touching to imagine a mother dinosaur delicately moving her newly hatched young to a safe location using her sense of touch.

Given their approach to hunting, and 400 million years or so of evolution, we should not be surprised that spiders have a highly refined tactile sense. Many species, from the humble garden spider to giant orb-weavers, build beautiful and elaborate silk structures strung across vegetation to capture flying prey. Silk is one of nature's almost miraculous substances—immensely strong, yet flexible and light. Its stickiness also serves to trap prey that fly into it. It forms something of an extension of the spider's sensory abilities too, allowing the occupant to deduce much information about what's landed in its trap. By building webs, or even just laying silk lines, a spider can extend its sensory world well beyond its own body. In most web-building spiders, individ-uals rely almost entirely on their mechanical sense to detect prey and to respond to potential mates, though this is not to say that other things like smell and vision are not used as well. Some spiders, particularly jumping spiders, have an extraordinary sense of sight that guides much of their life, but for many spiders, tactile information is key to their success. In this, their abilities are impressive and varied.

In orb web spiders, which build roughly circular silk webs, it has been suggested that some species could detect objects as small as a 0.05 mg piece of wire—about the weight of a single human eyelash—landing on a silk strand.[14] Different prey items produce particular vibration frequencies on the silk web strands, and spiders can often distinguish between these (though not always perfectly, since the frequencies of different stimuli often overlap). Other salient patterns of vibration can be decoded too, such as those of an approaching mate or predator. Of course, not all spiders hunt in this way, by building large webs. In fishing spiders (*Dolomedes triton* and *D. fimbriatus*), for example, prey capture behaviour can be elicited by both surface waves and by air-borne vibrations. These remarkable spiders sit on the water surface, using their legs to detect ripples and movement, at which point they dash across the water, using surface tension, to capture prey. Other spiders, such as many wolf spiders, rely greatly on vibrations transmitted through the ground, to communicate with one another, for instance in mating, and for orientation towards threats and prey.

Spiders are equipped with a variety of sensors to detect mechanical information, including fine hairs sensitive to wind movement and touch, and special organs called slit sensilla around the joints of legs that measure physical forces acting on the exoskeleton (Figure 26). It is these latter slit sensilla that we will focus on here. These organs have been described since at least 1890, and experiments in the 1950s showed that they are used in assessing mechanical information. Slit sense organs are found on a variety of arachnids, not just spiders, and they

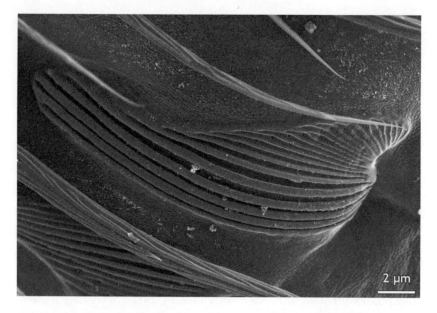

Figure 26. A scanning electron microscope photograph of the slit sense organs on the walking legs of a male spider, *Argiope bruennichi*. These organs respond to mechanical information, predominantly from the vibrations travelling through the ground or other substrates.

work by measuring strain on the exoskeleton coming from both external factors, including the vibrations produced by prey or mates, and internal factors, such as from muscle action or changes in haemolymph ('blood') pressure. Of these slit sensilla, one particular type has been very well studied: the 'lyriform' slit sense organ.

Mechanical forces acting on the exoskeleton manifest as deformation of the cuticle, and the spider can then measure where this deformation occurs, and the nature of it, by using its slit sensilla. One spider, the tiger wandering spider (*Cupiennius salei*) found in Mexico and other parts of Central America, has been particularly well studied by Friedrich Barth from the University of Vienna. This spider does not use a web to capture prey but hunts at night by sitting and waiting for prey to come near (like a cockroach or cricket), and hiding out in plants during the day. It is a large spider with a leg span of 10 cm (females are larger than males), and it is easy to breed in the laboratory, making it a valuable species for study. It is also well studied for its potent venom.

On the body of the wandering spider, especially on its legs, are around 3,500 slit sensilla. As the name suggests, these organs have elongated 'slits' where the cuticle is thin, and these respond to different levels of compression. With opening widths of around 2 μm, they are exquisitely sensitive, responding to compressions as small as 10^{-7} cm, enough to encode even the tiniest of vibrations that might travel through the substrate, even those made by the spider itself when walking.[15] When a vibration stimulus is detected, the combined information received from the eight different legs determines the angle by which the spider should turn to face the source, in particular by using information received from the fore legs. Often, the slit sensilla are arranged in groups (called compound organs), with up to nearly 30 of the sensilla sometimes occurring in the same area. An arrangement like this seems to allow organs that respond to different sensitivities, directions, and frequencies of vibrations, to be brought together to form a bigger picture of the tactile information being received. This is the case for the lyriform organ.

The extreme sensitivity of the wandering spider's slit sensilla mean that it is able to detect vibrations passing through a plant from a cockroach walking more than half a metre away. It's as if we were able to detect the vibrations made by a dog wandering around the opposite end of a football field. The spider can tell the difference between a prey animal walking and the effect of wind because wind vibrations on the plant tend to occur at lower frequencies and create very different patterns of movement. Correspondingly, when recording from the sensory nerve cells in a slit organ on the leg, the neurons fire most when the organ detects high-frequency prey movement vibrations, and not those associated with air movement.[16] The early stages of the sensory system are therefore already filtering out noise from the environment and focussing in on relevant stimuli.

Surprisingly, much less work has been done on how orb web-building spiders use their sense organs to detect prey and visitors. No doubt, they too can determine the frequency, intensity, and direction of vibrations and use this in mating, noticing intruders, in courtship, discerning prey from other objects, identifying prey types, and more besides. That said, research is not entirely missing and some of the earliest observations are both curious and creative. In 1880, one CV Boy published a letter in *Nature* entitled 'The influence of a tuning-fork on the garden spider'.[17] The work is just that. As Boy puts it: '...*while watching some spiders spinning their beautiful geometric webs, it occurred to me to try what effect a tuning-fork would have upon them.*' The spiders, it appears, were always very keen to run towards the fork when held vibrating against the web, taking it for a prey item. This certainly indicated that the spiders responded to vibrations on the silk threads. Boy also showed that they would only know the source of the vibrations when sitting in the middle of the web, so being in the centre of the structure appears key to measuring changes in vibrations coming along the radial threads from various directions.

There was a wait of almost a century before further studies revealed more details concerning the mechanisms of prey capture by orb-web spiders. Some of this work focussed particularly on the sensory function of the hairs found on spider legs, whereas other research investigated the general capture and response behaviour to prey in webs. For example, spiders such as a South American species of golden silk spider, *Nephila clavipes* (Plate 7), clearly show defined responses to prey of various sizes and levels of activity and can use vibrations to make assessments of prey type in the web and how to respond with predatory behaviour towards it. The golden silk spider is an impressive beast; orange-brown in colour and large in size, it builds a web made of golden silk strands. The spider sits in the middle of the web waiting for prey to be intercepted. When that happens, responses include either directly running to the prey item, or plucking the web to acquire further information.

In the early 1980s, researchers began to investigate how vibrations travel through spider webs. Using a technique called laser vibrometry, studies measured the transmission of vibrations through different parts of the web of species such as *Nuctenea sclopetaria*, a small brown spider that often builds webs near lights.[18] Laser vibrometry involves shining a laser beam at an object and measuring features such as the frequency and types of movement when the object is stimulated. The silk strands of a spider web could in principle vibrate over a range of frequencies, and do so in different patterns. A movement could cause a web to vibrate up and down, or side to side, or transmit vibrations from one end to another. By stimulating web strands in different ways in areas where prey might be caught, measurements can be taken of the vibrations that are transmitted to where a spider would sit in the centre of the web. This reveals that some types of vibrations, particularly longitudinal ones where the vibrations move along the axis of the radial threads, are transmitted very effectively; vibrations in other planes less so. This matters because longitudinal vibrations, unlike other types, seem to give good information about the direction of a vibration source on the web if the spider compares vibration information from different places, such as those arriving at different legs monitoring various regions of the web. In spider webs, just as Boy had suggested, the radial threads appear to be important in judging where prey is located, and certain types of vibration are particularly revealing for this purpose. These vibrations likely correspond to the sensitivity of the spider's sensilla organs.

Webs are wonderful structures at transmitting information about prey type and location, but how do spiders utilize this knowledge? Scientists have found that when vibrations occur in the web from actively moving prey, the spider can react to them in 0.1 s, turning rapidly towards the source. The spider is most sensitive to high-frequency vibrations. To locate immobile prey in the web, the spider tests the threads by vibrating them with its forelegs, and then measuring the change in vibration 'echoes' that return, to gather information about anything caught in the web. There is still much work to be done on how spiders analyse web vibrations, but as we would expect, they are adept at it. If you've ever put your finger on the strand of a spider's web and gently twanged it, waiting for the spider to run towards you and then wondered why it was not fooled, it's because it probably wasn't. It knew that whatever was on its web, it was not prey.

Many spiders also use vibration signals in courtship and mating behaviour. Although the mating interactions of the wandering spider C. salei are only brief encounters, they have a well-developed communication routine. The female lays a dragline laden with pheromones over plants that entices the male when he discovers it. The male then produces vibrations with various parts of his body that travel through the plant leaves, and the female responds with her own signals within about a second. Distinct and repeating patterns in the signal serve to distinguish them from those made by a walking prey animal. So in this case, it is not simply the frequency of the vibrations that enables the slit organs of the spiders to identify their origin, but rather the pattern of vibrations over time, and these patterns are species-specific.[19] Furthermore, the female's sensory organ, specifically the lyriform organ found on the leg, is particularly sensitive to the male's vibration signals. Recordings from the organs show that they only respond to the presence of several repeated components (syllables) of the male's display. During courtship, male and female wandering spiders have even been shown to communicate with vibrations at up to 3.8 m away on banana plants in Costa Rica. Imagine putting your finger on a tree and not only measuring the presence of a spider's vibrations from nearly 4 m away, but also being able to tell what it is, and what it is saying, and you can appreciate how refined is the spider's mechanical sense.

Mate choice in many spiders relies on vibration signals, and here much attention has been given to *Schizocosa* wolf spiders, a group of spiders occurring widely around the world. Various *Schizocosa* species use vibration signals from the male during courtship. A male complements these vibration signals by adding visual components to his display in the form of dark hair tufts on his legs. Scientists have shown that females primarily attend to the substrate vibrations and tend to ignore the visual components. That is, unless the vibration signals are not effectively transmitted through the substrate, in which case the visual signals come into play. Here, the environment also plays a key role in mating because the type of substrate has a major influence on the transmission of different vibration frequencies. Studies of one species, *Schizocosa ocreata*, have illustrates how leaf litter transmits vibration signals much better than do rocks and soil, which are very hard. Consequently, mating success of displaying males is higher on leaf litter, and when given the choice individuals prefer to court females on leaf litter than on other substrates. When males are forced to display to females on soil or rocks, they compensate by using more visual components.

In web-building spiders, vibration signals are also often key to mating success. In fact, not only do they provide a means for the male to signal his intentions to the female, but they are also crucial to the male in avoiding being eaten. Frequently, the male is much smaller than the female: this sexual dimorphism is extreme in some giant orb-web *Nephila* spiders, where the male can be a couple of centimetres in diameter but the female over 20 cm. Were he to clumsily blunder into a female's web he would likely trigger an instinctive predatory attack and end up as dinner. Instead, when entering a web the male must gently and characteristically twang the silk threads in such a way as to identify himself as a possible mate and subdue the female's attack response.

The mechanical senses of animals, which enable them to respond to vibrations, touch, and pressure, underpin many of the tasks a great variety of species have to perform. From the lateral line controlling shoaling movements in fish, leg sensors for finding and analysing prey and mates in spiders, to dome-like sensors on crocodiles and moles, a mechanical sense gives refined information about the world. It provides high precision and in rapid time. Unlike senses such as hearing and vision, which involve a small number of organs (eyes, ears) located in just a few body regions (e.g. the head), the mechanical sense is often distributed over large parts of the body. The lateral line runs along the side of a fish and the slit sense organs occur all over the legs of spiders. In humans, our sense of touch covers much of our body. This wide distribution of sensors for touch inherently enables animals to form a picture of their environment around them. Yet there are also areas and organs that tend to be highly sophisticated and that become centres of special importance, including our fingertips, or the teeth and whiskers of many rodents. More often than not, these reflect tasks and activities that the animal must perform in its environment and way of life to prosper. As with some other senses, notably hearing, understanding the sense of touch in animals, especially mammals, has greatly facilitated our understanding of how the brain processes sensory information to guide behaviours quickly and efficiently. In Chapter 6 we move on to what is perhaps the most widespread sense of all, one that does not rely on stimuli operating in wavelengths and frequencies, but rather in analysing discrete chemical packages of information: the senses of taste and, particularly, smell.

SMELLING
IN STEREO

To describe an animal as a 'swimming tongue' may seem strange, but it is a phrase that has often been applied to one creature that excels in detecting chemicals in its environment: the catfish. The description is apt because while catfish have many taste receptors in a cavity at the back of their mouth, for some species their whole external body surface is covered with taste buds. The channel catfish (*Ictalurus punctatus*) is an extremely common species in North America, and widely reared in aquaculture. In fact, it is fished by millions of anglers each year. It lives in murky freshwater, where vision is of limited value. Its taste buds respond to low concentrations of different amino acids in the water, valuable in detecting food in the environment, and when the buds are activated they often trigger swallowing responses. As expected, a range of genes encoding for taste receptors are found in the catfish and those receptors underpin its ability. The taste receptors are even found in the gills and especially on the whisker-like barbels. Such a refined taste is also a function of the sheer number of taste receptors the animal has: between 100,000 and 150,000 or more, compared to just

5,000–10,000 in humans. This enormous number, and the ability to detect different flavours and variations in their concentration all over the catfish's body, enables it to analyse the presence of food, where the food occurs, and what it is.

The chemical receptors possessed by some catfish can even enable them to detect minor changes in water chemistry that indicate the location of prey, in this case worms. The marine Japanese sea catfish (*Plotosus japonicus*) is able to measure differences in water pH of less than 0.1 using sensory cells on its barbels.[1] Such a pH change reveals the presence of worms because as the worms respire they release CO_2, and this causes tiny levels of acidification of the water around them. So, while harbour seals can detect the breathing of a buried flatfish based on minute water currents, the sea catfish detects the breathing of tiny worms via subtle chemical changes in the surrounding water. The use of chemical information by animals can sometimes be so refined as to be scarcely believable.

Detection of chemicals is perhaps the most widespread sense in nature, in part because chemical information is everywhere and very diverse in the information it can provide. A sense of taste is one way of acquiring information from chemicals in the world. The other major route to do so is through smell. Sometimes it is convenient to think of the ability to smell, termed olfaction, as the detection of chemical information when the animal is not in direct contact with the source of the odours. The odour molecules have to travel through a medium such as air or water to contact the receptors of the animal. In taste, the sensory organ is often in physical contact with the object or substance being tasted. This distinction between taste and smell may be convenient but it is not really accurate, in that the two processes truly differ in the specific types of receptor cells that are involved (mainly taste receptors in taste, olfactory receptors in smell) and the way that these are wired to and in the brain. Nonetheless, both smell and taste are heavily linked to an animal's ecology, as we found out already with penguins living in cold climates. Our senses of taste and smell allow us to detect and distinguish a variety of foods, potently harmful substances, and even influence our social behaviour, but our abilities clearly cannot compare with many other animals. This is especially true for our sense of smell, and this sense is of vast significance for many creatures across the natural world because it is used in such a wide range of tasks. We therefore focus on smell in this chapter.

The sense of smell operates somewhat differently to many of the other senses, in large part due to the type of information involved and how it is transmitted. Unlike light, sound, and mechanical information, which we can often describe as waves of different frequencies and intensities, odour molecules are more akin to little packets of information occurring in discrete molecules. They have to move through a medium, being pushed along in water or air currents or gradually diffusing along con- centration gradients. And the nature of odour molecules means that they tend to move relatively slowly and can be affected by things such as water currents and wind. On the other hand, unlike light, for example, they can pass around physical objects, and because such a vast range of odour molecules exist, the diversity of information smell provides is huge. Olfaction tends to transmit information fairly slowly over a distance, but it can be highly accurate. Just as with the other senses, smell is used for many things, from finding food, judging relatedness and kin, locating and assessing potential mates, marking and defending territories, to much more. The patchy nature of odour molecules in the environment means that animals often have to use certain behaviours to make full use of the information available. A male moth searching for a female performs a characteristic zigzag flight pattern as it moves back and forth into and across odour plumes. Some birds seem to do this too, and it enables the animal to track and orientate in the direction of the odour flow and find the source (a mate or food). Without this behaviour they would often lose the odour trail.

Understanding how olfaction works enables us to appreciate the ways that it guides behaviour, and how animals encode information about their environment. A particularly useful place to start is with insects, since these have been very well studied and in many regards are quite representative of how olfaction broadly works in nature. Then, we can move onto some of the masters of olfaction, starting with ants.

Located on the antennae of many insects are the main sensory receptors. These appendages can be wildly differing in form across species, and even between males and females. Male moths of a variety of species have massively elaborated antennae that they use to detect the mating pheromones of females (Plate 8). Otherwise, these structures house a range of other receptors for encoding aspects of the world from temperature and humidity through to pressure. The form of the antennae very much reflects the life history of the bearer.

In insect olfaction, the organs in which the receptors are housed are the olfactory sensilla, resembling miniature nipples projecting from the body cuticle. Around the edges of a sensillum are little holes or pores where the odour molecules can enter and then diffuse into the liquid inside. Within the liquid are the ends of nerve cells, called olfactory sensory neurons (OSNs), on which there are receptor sites that have an affinity for certain kinds of odour molecule. When an odour molecule of the right configuration makes contact with a corresponding receptor site a nerve impulse is triggered. We can think of this as a lock and key mechanism. Some receptor sites are very specific in structure or shape, others less so. Imagine a receptor site of a given shape, say a triangle, that can only be triggered by an odour molecule of exactly the shape that fits it. Other sites may be more like a semi-circular pit that is triggered by any molecule that is approximately circular and fits inside neatly enough to make contact. A feature of olfaction is that each nerve cell has just one type of receptor site, though the form of the receptor site may be either very specific or rather broad in affinity. This means that each nerve cell only ever responds to one type or group of odour molecules (or even just one specific molecule, if it is very picky), with other OSNs responding to different sets of odour molecules.

The types of OSN that animals possess reflects their ecology. Odours are of crucial importance in many aspects of the lives of fruit flies. In *Drosophila*, one of the most common model lab animals, there are roughly fifty types of OSN, and we can group these into three categories that generally occur in different sensilla structures.[2] One group responds to pheromones; a second responds to food odours and CO_2; while a third group responds to a range of things including food and water. Some of these OSNs are highly specific, whereas others respond to as many as 30 per cent of all the odour molecules a fly may encounter. Unsurprisingly, given its association with fruit, two-thirds of all the fly's receptors are activated to some extent by odours found in fruit. This is not to say that fruit flies do not use smell in many other ways; in fact, much of their communication is chemical based, including mating pheromones. These attract individuals of the same species and can, for example, convey information regarding mate identity. Chemical information also has a major role in determining egg-laying sites, and flies may even use chemical information to hide or camouflage the smell of their eggs in the environment to prevent them from being cannibalized by larvae that have hatched already. As such, in the fruit fly, there are groups of neurons that encode their main food, and ones that encode smells that are key to mating and communication. As an aside, a sense of taste in flies is also important. While our taste receptors are packed onto an organ in the mouth, the tongue, flies do things rather differently: they have taste receptors on their feet. This makes perfect sense since a fly benefits from tasting the substances that it lands on. In fruit flies, taste receptors on the feet respond to chemicals such as sweet substances; and, when triggered, nerve systems cause the animal to pause rather than to fly away. Likewise, when house flies land on and then walk all over our food, they are checking what it tastes like.

Returning to smell, so far so good, but a problem that arises in olfaction is that a vast range of odour molecules exists in nature—how can all of these be encoded? There is a limit to the number of OSNs that any animal can feasibly have, along with limits to the amount of neural and brain investment that can be devoted to processing all the information gathered. An upper limit therefore constrains how many different types of olfactory receptors can be packed into a sense of smell. Somehow, a comparatively small number of OSNs must enable the animal to discriminate the vast range of smells it may encounter.

The answer to the problem lies in how the information from the sensory cells is processed. In insects, the neurons send signals to an area of the brain called the antennal lobe and to a series of structures called glomeruli. Crucially, all the OSNs of the same type pass information to the same glomerulus, whereas those with different receptors are linked to other dedicated glomeruli. What happens next is that the insect olfactory system analyses patterns of activity across all these glomeruli. Each pattern of overall activity corresponds to a given substance type. Imagine three glomeruli, A–C, with A strongly activated, B intermediately activated, and C not activated at all; this might encode a specific fruit odour. By contrast, if all three are intermediately activated, then it encodes something else, perhaps an odour from rotting meat. Now imagine there are 50–300 glomeruli, the typical range in many invertebrates, and the number of different activity patterns from the possible combinations is almost endless; hence the range of smells they can encode is massive. This mechanism for encoding a multitude of odours has sometimes been called an 'across-fibre pattern' because the degree of activity across different structures tells the brain what substances are present. Processed information from these glomeruli next passes on to higher centres of the brain, which combine information with other senses and converts them into, for example, motor patterns controlling flight or other such actions (Figure 27). These higher centres also control learning behaviour.

The example of an across-fibre pattern is, as is often the case in biology, simplified to an extent, yet it does a good job of generally outlining how animals identify odours. There are other ways of processing too, and the olfactory system sometimes treats smells that are of great importance rather differently. Ultimately, evolution is all about passing on your genes, and to many animals this means finding a mate. Pheromones are chemical signals that are stereotyped and specific to each species, and are a key way by which many animals locate and attract suitable partners. In the antennal lobe, where the glomeruli occur, there is also sometimes a larger structure called a macroglomerulus, dedicated to processing information about pheromones. This does not get compared in the same way with patterns of activity across the

Figure 27. The Polyphemus moth (*Antheraea polyphemus*), like many moth species, shows considerable differences in the size and structure of the antennae between males and females. Males, as shown here, tend to have larger and more elaborate antennae to detect female sex pheromones.

other glomeruli but rather forms its own direct pathway, some-times called a 'labelled line'. This arrangement essentially reflects the fact that the receptors for pheromones are very specific to those chemicals, or components of them. The macroglomeruli are large because there are a lot of receptors dedicated to process-ing pheromone components from which they receive informa-tion. In the American cockroach (*Periplaneta americana*) there are two specialist OSNs used for responding to two components of female pheromones: periplanone A and B.[3] Given that it is the males that have to respond to the pheromones, only they have a macroglomerulus for processing this information. As with many other senses, within the olfactory system, stimuli that are of high importance can be processed with greater investment than other odours the animal may receive.

When it comes to social behaviour, ants are among the most remarkable creatures on the planet. These hugely successful insects dominate many ecosystems around the world, and are among the most numerous organisms on the planet, in terms of both numbers of individuals and numbers of different species. Highly sophisticated social behaviour is a critical element of their success. Ants live in colonies, usually with one or a few queens that control the workers, with the latter carrying out the day-to-day activities for the colony. The queen is charged with keeping control and laying eggs to maintain the worker population. Worker ants (always female) take care of cleaning the nest, raising the queen's brood, foraging for food, and protecting their home from intruders and attack. The diversity in ants is staggering. Take, for example, the army ants, among which the workers are anatomically highly distinct. Individuals of some 'castes' look like they are beefed up on steroids and are equipped with massive jaws for defence and fighting, whereas other ants are quick and streamlined for foraging, even standing on the backs of other ants to ward off parasitic flies. Many other species lack such marked distinctions—in fact the workers all look much the same—but they still show strong divisions of labour in the main tasks each ant undertakes in the colony. Divisions of labour among workers, with some focusing more on colony defence and others on foraging, partly enables ants and other social insects, such as bees and wasps, to be highly efficient. Ant colonies also vary greatly in the ways in which they function, from the millions of ants in an army ant colony, which forms a nest that moves around in the forest each day, to the sizeable and impressive underground systems of leaf-cutter ants with their 'farms' of cultivated fungi. Not all colonies are vast; some species of ant have just a few hundred workers and brood, and colonies that fit entirely within a single acorn.

The key to ants telling each other apart—odour—has been appreciated since the early 1900s. However, it is perhaps especially in the last 15 years or so that substantial progress has enhanced our understanding of exactly how ants use odour to recognize other individuals or colonies. The body of each ant is marked by a combination of chemicals called 'cuticular hydrocarbons' or 'CHCs'. These are synthesized by the ant itself through different glands, and also acquired from physically touching the nest environment and other ants, such that an individual becomes bathed in these chemicals. In their classic synthesis, *Journey to the Ants*, Bert Hölldobler and Edward Wilson, two renowned experts, described ants as 'walking chemical factories'.[4] CHCs are specific to both the species and the individual colony itself, and sometimes in small colonies they might even be used to recognize individual ants too, though this is by no means well established. CHCs can, however, demark the sex, age, and caste of ants in a nest. When interacting, ants typically touch one another with their antennae, allowing them to better assess their chemical profiles.

The chemical information is critical to ants for many tasks: queens release pheromones to suppress workers from breeding (the pheromones themselves are sometimes CHCs); workers often recruit others to food sources via pheromone trails; and, when the colony is under attack, alarm pheromones are produced and can act as a call to arms for the workers. While CHCs differ between colonies, the queen pheromone is often superimposed on top of these. CHC profiles that differ between individual colonies and that are used for recognition of nest mates versus intruders generally need to be learnt, whereas pheromones produce more stereotyped responses that the ants instinctively respond to.[5]

Dominance is frequently a key element of socially living creatures. In some, for instance primates such as macaques, a strict hierarchy of dominance rank is maintained, whereas in other animals there might be one main dominant specimen who rules over all the workers, which are of much the same status. In ants, chemical information enables dominant individuals to express control over others. In most ant species, a dedicated queen ant, or sometimes several of them, exerts control over the workers. Yet there are a few exceptions, where the nests are run by the worker ants themselves. One such case is the black Brazilian ant, *Dinoponera quadriceps*. Colonies nest at the base of trees in the forest and are ruled by a dominant alpha female. Nests are small, with around eighty workers, and the alpha female is a worker that figuratively sits on top of a group of another four to five relatively dominant workers, one of whom would replace her should she die.[6] Sometimes, the alpha female is challenged by another worker just below her in the pecking order. When that happens, they fight intermittently and the alpha female wipes her sting on the challenger. Following this, the challenger is frequently punished by being held down by the other workers, sometimes for several days, and as a result she loses her high rank, or may even be killed. In marking the challenger, the ruling female applies chemicals from a special gland that frequently plays a key role in ant communication, a Dufour's gland; it is these pheromones that induce the other workers to supress the challenger. Dufour's gland pheromones are well known to produce a range of so-called 'propaganda substances' in ants, which can play a role in queens taking over other colonies, and in worker manipulation. In some ants, the substances even aid in attacks by workers on other colonies by making the opposing colony's workers turn on their own colony and attack themselves.

Advances in scientific techniques have provided some stunning demonstrations of just how critical odour is to ants, including in a recent study on the Indian jumping ant (*Harpegnathos saltator*).[7] The name 'jumping ant' is apt, since individuals can literally jump up to a few millimetres high. Workers are red and black with long mandibles and they are active hunters that venture out mostly in the mornings to look for prey. By using new methods to 'knock out' different genes in the ants that would normally encode for a wide range of odour receptors, the scientists created specimens that could no longer detect pheromones effectively. The results were marked. In these ants that had lost their odour receptors, the part of the brain that processes odour information, the antennal lobe, was observed to be substantially underdeveloped. The mutant ants became 'wanderers' spending much more time outside the nest, as if they could not sense key chemical signals in the nest or compare smells inside with those outside the nest environment. They were also less able to detect potential prey, such as crickets, or bring food back to the nest, and showed less interaction and contest behaviour with their nest mates. Mutants also showed a reduction in mating behaviour, amounting to reduced grooming of others, presumably because they could not detect key sex pheromones. A loss of smell renders the ants incapable of performing a wide range of critical tasks and they become almost like zombies.

Regardless of the great diversity of ways ants live their lives, one commonality among them is the need to recognize individuals from the same versus other colonies. Ants are highly territorial, ferociously defending their nests and resources from individuals from other colonies and from other species. Worker ants are typically all offspring originating from one or a few queens, so they are sisters, and driven to protect their kin and the brood. An ant from another nest found in the vicinity is typically attacked and shown no mercy. This is not without reason, because most ant nests are under constant risk of attack from ants from nearby colonies and species with which they compete, as well as a whole host of invaders that pretend to be ants in order to infiltrate the nest and live off the colony resources. Being able to tell apart individuals from their own colony from those of other nests is critical for workers, and this largely comes down to smell, though they do also use vibration and sound cues too.

The recognition of nest mates is based on refined detection of a number of CHC compounds. One of the ways by which this has been tested is to present ants from different colonies with objects, such as simple glass beads, which have been marked with the chemical profiles of different nests, and then changing specific aspects of the chemical make-up to determine how the ants respond. The ants, being so driven by chemicals, will show different behaviour (aggression or grooming) towards beads that either lack key chemical profiles of their own colony or have components from a different colony.

How do ants achieve such refined abilities to tell odours apart? CHCs are highly diverse and complex so this is not straightforward. In fact, how ants achieve this is not fully understood, and given the diversity of chemicals involved, not easy to test either. Again, it was at the turn of the twentieth century that the antennae were implicated in this ability, but only far later did we learn more. For example, in 2005, Mamiko Ozaki, from the Kyoto Institute of Technology in Japan, and various colleagues found a receptor organ (a sensillum) on the antennae of the Japanese carpenter ant (*Camponotus japonicus*) that is used in identifying nest mates.[8] The receptor seems to respond to the smell of non-nest mates. Ozaki and her colleagues treated glass beads with the

CHCs of ants from either their own or other colonies and showed that ants were much more likely to be aggressive towards beads with foreign chemicals. Next, they made nerve recordings and found that the sensillum on the antennae responded to non-nest-mate CHCs, but not to the CHCs of nest mates. Interestingly, the sensillum contained many (around 200) potential receptor neurons, and Ozaki and colleagues suggested that each of these may respond to different components or molecules of the complex CHC profile. So, the ants may encode foreign smells, rather than constantly responding to the odours of their own colony.

Thanks to advances in molecular biology, we now know that ants have a staggering diversity of receptors for detecting odours compared to many other insects. In addition, their repertoire seems to have been very flexible over evolutionary time, with gains and losses in these receptors, probably linked to ant species diversifying and moving into new modes of life and habitats. In ants and other highly social insects like bees and wasps, the workers are female whereas males tend to only be produced to leave the colony and mate with new emerging queens, which then go on to form new societies. Thus, the demands and functions on males and females can be very different. And as we might expect, strong differences can arise in the odour receptors used by male and female ants. As with most senses, ecology dictates the make-up of the olfactory system.

Somehow, the ant's olfactory system can detect both general odours like food and more specific ones like pheromones, alongside measuring complex CHCs used in nest mate recognition. Ants like the Florida carpenter ant (*Camponotus floridanus*), for example, have sensilla found on workers with multiple sensory neurons that respond to a broad range of hydrocarbons, spanning odours on nest mates through to queen pheromones.[9] By contrast, other olfactory receptors in ants are more narrowly tuned to certain odour molecules. Ants may apply a spectrum of processing, similar to that of the fruit fly, which ranges from across-fibre patterns to something more like a labelled line, to respond to both general odours and those of particular importance. The task of discriminating between CHCs of ants may be so complex that the whole odour system is needed, and then at some stage the information about what type of stimulus exists is segregated. Large combinations of numerous olfactory receptors, each variously tuned to specific odours, may allow the workers to differentiate between the highly complex CHCs and pheromones that dictate their social lives. What we need to understand better is how processing, early or later on in the brain, enables ants to classify these numerous odours and recognize them, and how they are then able to take appropriate behavioural action.

While smell is often thought to be the most widespread of all senses in nature, our understanding of how it works has sometimes lagged behind that of other senses such as vision and hearing. When it comes to mammals, major breakthroughs occurred from the early 1990s, following the identification of genes that encode for the olfactory receptors found on sensory cells in 1991. This discovery, by Linda Buck and Richard Axel, brought them a Nobel Prize some years later.[10] The findings enabled scientists to characterize the nature of olfactory receptors that respond to odour molecules. And since then, research sequencing the genomes of animals, including humans, rodents, dogs, and other species, has further facilitated our understanding of the genetics involved in the evolution of smell. All in all, smell is so important to many mammalian species that the genes that underpin it are thought to be the largest family of genes in the whole mammalian genome.

One group of animals that immediately spring to mind when thinking of creatures with excellent smell are dogs. There is no doubt that their sense of smell has contributed to their long working relationship with humans, from help in hunting to search and rescue, spanning the 15,000 or so years since their domestication from wolves. Our knowledge of dog olfaction has developed considerably in the past decade, in part thanks to detailed work on the genetics and physiology behind its workings.

Chemical communication is clearly extremely important in dogs. It is used in a whole manner of interactions in wolves, not least in marking and detecting territory boundaries with urine. Anyone who has spent even a few moments around dogs in a park cannot help but notice how much time dogs spend sniffing the ground, around objects like trees, scent marking numerous times, and, of course, smelling each other. After being domesticated for so long, dogs are also extremely good at reading humans, and this has clearly been a valuable trait for breeders in producing a variety of working and companion dogs. Many breeds are well attuned to respond to our gestures and voice, and our smell, with the latter even seeming to include our emotional state. For example, a so-called 'smell of fear' may seem far-fetched, but there may in fact be some grounds for this in canine odour detection. When humans are fearful or under stress our body chemistry changes, and as a result so can the odours we produce. Dogs may be able to detect these changes in odour associated with condition or emotion.

For some time a belief has spread that dogs can detect changes in the odour of humans suffering from disease even though the person may not yet be aware of it, potentially enabling earlier detection and treatment. While this has not always been easy to demonstrate scientifically, dogs do seem capable of detecting the presence of diseases such as skin and prostate cancer in patients, with a high degree of accuracy. More widely, canines are routinely used to detect drugs and explosives (something we will return to in Chapter 8). They have even recently been used to track tigers in the Russian far east by cuing in to odour from scats, and this may be valuable in estimating numbers of these endangered wild cats, and even tracking specific individuals. Clearly, then, dogs must have sophisticated apparatus for discerning smells.

Just as in other mammals, air enters the nose of a dog and comes into contact with an area called the olfactory epithelium. This is a region of membranes lined with mucus that sits on top of a bony structure (Figure 28). Being highly folded on itself, the epithelium creates a large surface area for trapping odour molecules that are then detected by the olfactory sensory cells. The surface area of the membrane in dogs is much greater than in humans, occupying 200 cm² in a German shepherd and 67 cm² in a cocker spaniel, compared to just 5 cm² in humans; and the total number of receptor cells in dogs is thought to be over 100

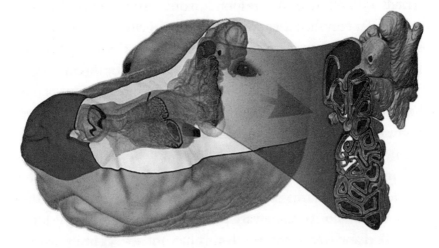

Figure 28. In dogs, odours enter the nasal cavity, which houses the olfactory epithelium and the olfactory sensory cells. The membranes of the epithelium are highly folded, greatly increasing the surface area such that greater numbers and more types of receptor cells for analysing smells can be housed.

million (more in some breeds), compared to estimates of around 5 million in humans.[11]

Research characterizing the processing and genetics of olfactory receptors in dogs began in the 1990s and took important strides forward in the early 2000s. Olfactory genes in both dogs and humans seem to have evolved by gene duplication from our ancestral mammal, and the gene repertoire has expanded much more in dogs than in humans. In fact, dogs have something like twice as many functional olfactory genes than humans (856 versus 391). The label of functional or 'active' genes is important here because across mammals there are high numbers of so-called pseudogenes—genes that are no longer operational—and this varies with species. Humans have lots of olfactory pseudogenes, in fact more than any other land mammal tested. Smell has apparently become less important to us over our evolution and some of our genes that played a role in smell have consequently drifted, with mutations building up that have rendered them defunct. In contrast, their large numbers of active genes should enable dogs to have a refined ability for discriminating among odours, though numbers of genes and receptors do not alone translate into a 'better' ability to smell. Dogs have fewer active olfactory genes than mice or elephants, yet this doesn't necessarily mean that an elephant has a better sense of smell.

Dogs also use a second area for detecting odours, a vomeronasal organ, present in a nasal cavity above the roof of the mouth. Used in characterizing smells, such as certain pheromones for communication, it comprises receptors that are different to those located elsewhere. However, the vomeronasal organ does not seem to be of major importance in dogs, despite their retaining at least some functional genes for these receptors. As such, this aspect of dog olfaction may be useful but not key to their exceptional abilities. The information from odours acquired by the olfactory epithelium and the vomeronasal organ are processed separately en route to the brain. Similarly to insects, nerve cells that stem from the sensory cells that respond to the same odours travel to the same areas (glomeruli) for later processing. Nerve cells that process other odours pass to different locations. Also akin to insects, this means that specific smells and substances can be encoded by general patterns of activity across multiple glomeruli.

To make the most of their sense of smell there are a few other neat tricks that dogs possess. For one thing, dogs vary their rate of sniffing depending on how tricky the task is, and sniffing odours is a very different mechanism to simple breathing. In sniffing, air makes prolonged contact with sensory cells and receptor areas, whereas in breathing for gaining oxygen the air goes much more directly to the lungs. Impressively, dogs have a system of air flow in which the inhaled air takes a different pathway to exhaled air. By doing this they separate inhaled fresh air, which carries new odour information, from air that they have already sampled, allowing the newly inhaled air and the odours contained to come into contact with the sensory cells for longer.[12] Furthermore, the noses of dogs are set up anatomically in such a way that air that flows to each nostril originates from different areas outside. Each nostril samples a different part of the environment, potentially allowing the dog to determine the location of an odour source and the direction to it based on bilateral information, something we will return to very soon.

Another remarkable feature of canine (and, in fact, of many if not most mammals) olfactory receptors is that they are constantly replaced with new cells over a period of weeks and months. Even more amazing is that the replacement cells do not have to be sensitive to the same odours as the preceding ones, meaning that the sense of smell is plastic and can respond to changes in the environment over time. So, for example, if a dog is found in an environment where some types of odour predominate, or they are trained to locate or discriminate specific smells, the sensory system can adapt and become better with experience at detecting the key odours. There are fascinating parallels here with both vision and hearing, in which, as we have observed already, the sensitivity of the visual system to light can change with diet and habitat, and the ears of animals such as birds can regenerate new receptor hair cells.

Breeds of dog, of which there are some 400 or so, are well known to vary considerably in their senses. Some are widely acknowledged to have a superior smell (bloodhounds and various sniffer dogs, for example). Across breeds, there is much diversity in the specific combinations of versions of genes (alleles), and some genes are especially variable. Differences in these features across breeds may relate to their olfactory ability. As such, it is not specifically that some breeds have more genes for smell per se, but rather that they have versions of genes that afford them an advantage, and fewer pseudogenes. Other differences among breeds, not least the size of the olfactory epithelium, also contribute to the differences in smell capacities that exist. These comparisons among types of dog translate into better or worse performance, in at least some tasks.

High numbers of receptors and a large surface area for housing these in the nose, plus some neat tricks to create and analyse different air flows, allows for excellent canine odour detection and discrimination. However, in spite of this, there remains no easy way to judge just how good a dog's sense of smell actually is, either in absolute terms or compared to other animals. Undoubtedly it is extremely refined in many breeds, but part of the reason dogs are so effective also comes down to something else: their ability to learn to use their smell and be trained by humans. It has been suggested recently that cats actually have a superior sense of smell to dogs—if only we could train cats and use them for our purposes!

Our own sense of smell is often underestimated; on the whole it is in fact relatively good. That said, when it comes to making judgements regarding the intensity and source of odours, humans can only make relatively crude assessments. A far more advanced and refined sense of smell, including the ability to determine the location of an odour source precisely, occurs in other animals. Some, like dogs, do this by sensing in stereo—not dissimilar to the way in which we combine information from both eyes into one picture, or combine sounds from both ears into the same sensation of hearing. In smell, some animals compare information about odours arriving at their two nostrils in a very sophisticated way.

Most moles are not well regarded for their vision; they do not really need it in the subterranean environment where they live. But they tend to compensate for this with other senses, especially a good sense of touch—sensationally so in the case of the star-nosed mole—or with decent hearing. The eastern American mole (*Scalopus aquaticus*), however, lacks sensitivity in these senses as well. It is a mole widely found in the United States, making burrows and mole hills in woodland and meadows. Not only does it have tiny eyes but its sense of touch is much worse than that of other moles, and its hearing is mainly limited to sounds of low frequencies. But it makes up for these deficiencies with a supreme sense of smell. The eastern American mole is one of the several mammals that has been shown to smell in stereo.

The mole is fast and accurate at finding prey such as earthworms, taking just a few seconds to do so. Kenneth Catania, who we encountered in Chapter 5 through his work with the star-nosed mole, tested how it does this.[13] The moles repeatedly sniff as they move their nose, constantly sampling the area around them to detect and analyse smells. Catania showed that if you temporarily plug one of the mole's nostrils it will move in the wrong direction when heading to an odour source, especially in relatively close proximity to the source. Blocking nostrils is not seemingly a problem with long-distance odour detection because from afar the task is to broadly follow a somewhat dispersed pattern of odour molecules to the right general location. On the other hand, blocking a nostril is bad for pinpointing the exact location of odour sources over a few centimetres when the mole needs to make refined assessments in a small area. The mole, it seems, can separately encode odours from the left and right side of the body and use this to work out where exactly a food item is.

Rats, which also have an excellent smell, may use bilateral information too. The ability may even be useful for detecting predators, and in knowing where exactly predators are lurking. The findings in the stereo mole essentially parallel some of the features of sound detection, rather like the way in which owls zero in on hidden prey based on the noises they make. Owls combine information from both ears, and the brain makes comparisons in the timing and intensity of sounds that arrive at each ear in order to locate the position of a sound source in space. In the case of olfaction in the mole, blocking one nostril causes a mismatch in olfactory information arriving at each side compared to the real scenario, leading to the mole making mistakes in locating the position of the odour source. How widely stereo smell occurs is unclear. It likely exists in other mammals too, maybe even in humans to a certain extent, but without further experiments the eastern American mole is the stand out performer, for now at least. Beyond mammals, smelling in stereo exists elsewhere in vertebrates, perhaps most famously via the forked tongues of snakes. Rather than smelling directly with their tongues, the animals sample chemical information from the air and ground by collecting it onto either side of the fork. Next, the odour molecules are transferred into the mouth, and then onwards to organs in the roof of the mouth that do the actual smelling. By keeping the odours collected from each side of the tongue separate, the snake can smell in stereo.

Odour is key to the way in which many animals interact, and it is used in a staggering range of tasks. In mice it enables judging familiar individuals and even assessments of how closely related others are so that close kin do not mate together. Many animals, such as bears and wolves, use smells to mark territories, and smelling has the advantage over many other senses in that the odour molecules hang around after being deposited. This property is what enables ants to lay down odours in the form of trail pheromones, leading other workers to the site of food that a scout has discovered. Visual or auditory signals, such as avian territorial song or plumage displays, are far more transient. Instead, an animal such as a black bear, which has excellent smell, can mark out the boundaries of a territory by rubbing on trees or vegetation, and those scent markings will persist long after the territory owner has departed. Far from being a mammalian trait, odours are used in similar ways by invertebrates, such as with lobsters, where the odours from urine are used in competition and to enforce dominance hierarchies. More remarkably, odours can be used over extremely long distances. It is tempting to assume that smell operates in fairly close proximity, for instance over centimetres or metres, because that is rather within the limits of our ability in most cases. Yet a male moth can track down a female from her odour plume over hundreds of metres, sometimes even several kilometres. More impressive still, fish such as salmon use odour cues over hundreds of miles to travel back to their natal spawning grounds. Odour and a sense of smell give animals something that is often quite different from the other senses, a persistent source of information that can operate over very long distances and encode extremely high levels of accuracy in detecting, localizing, and discriminating objects and individuals. In Chapter 7 we come to another sense that, along with the electric sense, humans lack entirely, and one that also enables animals to travel long distances: the magnetic sense.

CHAPTER

7

HOMING TURTLES AND ANIMAL MAGNETISM

The accolade for the most extreme migration of any animal must surely go to the Arctic tern (*Sterna paradisaea*). During the northern summer it breeds in Europe, Asia, and North America throughout the Arctic and sub-Arctic regions, before heading south all the way down to the Antarctic for the southern summer. En route, birds originating from different northern locations pass through a variety of areas, as far apart as western Europe to New Zealand before reaching their destination. Those individuals that come from the Netherlands have been shown to follow the coasts of Europe and Africa, turning east at the bottom of Africa, before crossing the Indian Ocean, and then travelling onwards towards Australia, at which point they head south to eastern Antarctica (Wilkes Land). Far from heading directly south, they travel half way around the world as well. Estimates from some birds show that they can travel up to 90,000 km in these migrations.[1] To put that in perspective, combined over the lifetime of a bird, which can be 15–30 years, Arctic terns can travel the equivalent of more than three return trips to the moon. Young birds are quickly able to undertake the migrations; individuals from the UK and Canada have been found in Australia and southern Africa just a few months after fledging. These dramatic journeys enable the birds to experience two summers each year, and more daylight than perhaps any other animal.

Arctic terns are exceptional in the distances they cover, but their abilities can be matched in other ways by other species. Alpine swifts (*Tachymarptis melba*), for example, can spend six months in continuous flight while migrating, feeding, and even sleeping on the wing. Twice a year, bar-headed geese (*Anser indicus*) must travel across the extreme high ground of Tibet and the Himalayas, sometimes flying at up to and above 7,000 m when migrating, at altitudes where there is less than 10 per cent of the oxygen found at sea level. Climbers have even reported seeing geese flying over Mount Everest. Birds can also travel great distances in comparatively short time scales, including the bar-tailed godwit (*Limosa lapponica baueri*), which can travel 11,000 km over eight days of non-stop flight, journeying from Alaska to New Zealand without breaks to feed or rest. In songbirds, the northern wheatear (*Oenanthe oenanthe*), a bird of just 25 g, travels 14,500 km twice a year. During a one to three month migration, its movements take it over oceans (sometimes encompassing a path across 3,500 km of the Atlantic), deserts (the Sahara), and Arctic ice fields from places such as Alaska to areas of sub-Saharan Africa. Taking in these facts and figures is a challenge in itself, yet what is even more remarkable is that many birds return not just to the same geographic regions year after year, but to the same specific sites. They have a staggering ability to calculate where to head and when. In fact, in one study where researchers displaced migrating reed warblers (*Acrocephalus scirpaceus*) by as much as 1,000 km across Russia, the birds still corrected for this displacement and adopted a modified orientation to get to where they needed to end up.[2] Arriving at the planned destination matters, because these sites are frequently places that provide good opportunities for nesting and feeding, and returning to them each season is likely to bring much greater success than would be the case if the final stop was more haphazardly chosen. Birds, alongside a variety of other animals, use many sensory inputs to find their way, from the position of the sun, starlight patterns at night, even to odour cues. However, one of the chief senses they exploit is something completely alien to us: a magnetic sense.

While it may be completely alien to us, a magnetic sense is widespread in nature and allows a variety of animals to detect the Earth's geomagnetic field, and to use this for orientation and navigation over short and longer distances. Unlike other sensory types, the magnetic sense cannot be used in communication between individuals. In terms of vision, animals often produce extravagantly colourful displays for the benefit of potential mates, as threats against competitors, or to lure prey; think of the stunning plumage and dances of birds of paradise, or the bioluminescent light emitted by countless creatures of the deep sea. And in the case of hearing, too, animals ranging from cicadas to lemurs call to alert others or for courtship. However, equivalent uses cannot work with the magnetic sense because, unlike superheroes from science fiction films, animals cannot control or create magnetic fields of their own, or manipulate them in any way—at least, as far as we know. The uses of the magnetic sense must therefore be passive, mostly in enabling animals to move around their environment and navigate over long distances. There are downsides to exploiting this sense for orientation. Over time, the Earth's magnetic field can change and even flip polarity, and quirks in local geology can influence properties of the field. Conversely, unlike sunlight, magnetic information is fully available day and night.

Considering the nature of the Earth's magnetic field helps us to understand how animals use it. The field is believed to originate from movements of iron-rich fluids in the Earth's core. By analogy, it is as if the Earth had a giant bar magnet within. The magnetic field lines can be envisaged as 'leaving' the Earth at the south magnetic pole (close to but not identical with the geographic south pole), arcing round to become parallel with the Earth's surface at the (magnetic) equator, and then 're-entering' the Earth at the north magnetic pole. The angle of the field lines relative to the Earth's surface is called 'inclination' or dip, and it varies with latitude (Figure 29). If we were somehow able to perceive magnetic inclination and to walk from the south magnetic pole to the north magnetic pole, the angle of the field lines that we would encounter projecting out from the Earth's surface would change: initially they would be pointing vertically up, then horizontally as we pass the equator, and then vertically down at the far north. Using this information it would be possible to know whether we were heading towards or away from the equator or one of the poles.

Besides inclination, another piece of information that an animal could acquire is the intensity or strength of the magnetic field in a given location. On a broad scale, magnetic intensity tends to be strongest at poles and weakest at the equator, and this can be used for determining position. Intensity also varies over smaller spatial scales, including quite locally. Some animals may calculate another feature, the angle between magnetic north and true geographic north on a horizontal plane (sometimes called declination). This can also help with determining direction, rather like a compass. Superimposed on top of the larger-scale changes in the Earth's magnetic field are variations in field properties due

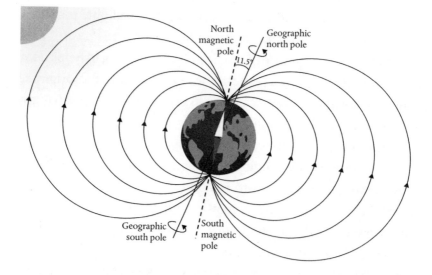

Figure 29. The Earth's magnetic field allows animals to measure a range of information, including intensity and inclination. In the latter, the magnetic field lines 'arc' out into space from the magnetic south pole, become horizontal relative to the Earth's surface around the equator, and then arc back down towards the magnetic north pole. An animal sensitive to these angles of inclination or 'dip' in the field lines could deduce whether they were travelling towards or away from the equator or poles.

to local anomalies, including geology and other factors. These variations can either cause 'noise' that interferes with valuable information or instead be useful if the anomalies correspond to a specific location that an animal might wish to find.

Animals might guide their movements based on magnetic cues in two main ways, though in reality, likely more in a continuum between the two. First, the animal could possess a magnetic compass. It could then head in a specific, consistent direction such as north or south, or 'poleward' or 'equatorward', potentially by using information from features such as inclination. Such use of magnetic compasses by animals is uncontroversial, but in order for the bearer to find a desired final location it needs information from the other senses too. A compass is rarely sufficient to arrive at a precise point in space because it will often not be accurate enough on a small scale, and animals can be moved off their trajectory by wind or water currents or other factors. Using input from other senses as well, perhaps starlight patterns or odour information, enables animals to refine their navigation routes.

A more sophisticated capacity that some animals may have is to use a magnetic map. By providing specific information about current location and destination, a map affords the ability to navigate by determining geographic position, rather like having a natural inbuilt GPS system. In the simplest sense of a map, individuals can simply work out their geographical position or location. This does not necessarily mean that they have any mental or internal representation of where they are specifically in space in the way that we might read a road map. A map may, for example, simply enable an animal to know that it next needs to move west or east to get home from its present location. There are in fact lots of different ways in which a magnetic map might be used, but we need not worry too much about that here, rather just note that the map enables an animal to compute location, instead of being only a source of directional information. Now that we broadly know how animals can use magnetic information, we can move on to our first animal of this chapter: the sea turtle.

How animals use magnetic information to get around is wonderfully illustrated by the much-studied sea turtles. Extensive work by Kenneth and Catherine Lohmann at the University of North Carolina, much of it undertaken with various colleagues, has sought to understand how loggerhead turtles (*Caretta caretta*) use a magnetic sense. Hatchlings emerge, mostly at night, and, as we noted earlier, use the light of the moon to guide them to the ocean, since light reflects more off the water surface than land. Once in the water, they swim out to sea, establishing the direction away from shore in part by orienting themselves into and swimming against the movement patterns of incoming waves. Magnetic information then becomes crucial once individuals have ventured beyond these shallow waters. Hatchlings emerging from beaches in North America move off the coast to the relatively warm waters of the Gulf Stream and the North Atlantic gyre, a circular current system within which they must remain for several years while developing. This oceanic current seems to provide the turtles with important feeding opportunities and relatively warm water, but to stay within it they must orientate themselves and swim in the correct directions, based on their current location. Over time, while growing and then after reaching maturity, some turtles will migrate hundreds, or even thousands, of miles between feeding and nesting sites.

Early in the 1990s, to test whether turtles respond and orientate based on magnetic cues, Kenneth Lohmann put hatchlings from Florida in a tank that was surrounded by electromagnetic coils, such that the magnetic field could either resemble natural conditions at the location or, by changing components of the field lines, somewhere different.[3] The turtles were tethered via a 'back pack' to a lever-arm that could track the direction in which they tried to swim for up to two hours, all in complete darkness. Lohmann found that when the field lines were reversed (shifted by 180 degrees), the turtles changed their orientation from an angle of 42 to 196 degrees (a shift of 154 degrees). This was clear evidence that the turtles could detect the Earth's magnetic field and orientate in line with it.

Further work tested the orientation behaviour of turtle hatchlings to different aspects of magnetic information including the angle of inclination (dip).[4] Since the angle that field lines intersect the Earth varies with distance north or south from the equator, by responding to inclination a turtle could deduce its latitude. In hatchlings especially, this may be valuable for staying in important ocean currents. Loggerhead hatchlings presented with inclination angles corresponding to locations where they had hatched orientated themselves in an eastward direction, consistent with their behaviour when migrating offshore. In comparison, when given inclination angles that matched those found at either the northern or southern points of the North Atlantic gyre (while keeping magnetic field intensity unchanged), the turtles swam in a southwest or northeast direction, respectively. Such behaviour apparently enables turtles to swim in the directions that would be required to stay in the ocean current. Failing to do this would mean leaving the gyre itself or taking one of the dividing currents that head away into colder northern waters. Since nesting beaches along the coast tend to vary in latitude, inclination may enable turtles to orientate in the right direction when they emerge.

The Lohmanns also showed that turtles detect and respond to the intensity of magnetic fields. Hatchlings which were presented with information corresponding to two different intensities found at two locations of the Atlantic Ocean differed in their swimming responses. This was potentially significant because to accurately determine location, as would be needed if they were using a magnetic map, animals must detect two aspects of the magnetic field that both differ in space, such as intensity and inclination. So, a decade after the initial work demonstrating the use of a magnetic sense in turtles, the Lohmanns and their colleagues further demonstrated how turtles stay in specific locations of the North Atlantic gyre. Such locations, and correspondingly different places along the gyre, vary in both magnetic field intensity and inclination. Deducing position in the gyre and staying in the system could be achieved by combining these components and then orientating appropriately. The researchers presented turtles with magnetic information in the form of intensity and inclination that matched three locations in the gyre. As expected, in each case the turtles swam in directions that were consistent with the mimicked location; their swimming directions matched those that would be needed in order to stay in the current at each place.

Since magnetic cues vary most with latitude but much less with longitude, it was uncertain whether navigation with a magnetic sense could accurately be used to deduce longitude. However, later discoveries showed that the turtles can calculate both latitude and longitude separately. When provided with magnetic cues corresponding to two locations on the North Atlantic gyre of the same latitude but of different longitude, turtles swam in different directions in each scenario. Using magnetic information to calculate both latitude and longitude should enable loggerhead turtles to have what's called a bi-coordinate magnetic map. One of the fascinating aspects of this work is that the turtles used had not migrated before—they had never been in the ocean current or experienced any of those locations in real life. This suggests that turtles inherit a magnetic map that allows them to calculate their position in the ocean and adjust their orientation appropriately so they can travel towards a specific goal. On the other hand, turtles could simply be pre-programmed to swim in a certain direction when faced with given magnetic information, without actually having to navigate in any way. For the loggerhead hatchlings we cannot be sure which is true, but given the sophistication of adults in being able to locate very fine-scale positions (e.g. specific islands), and their opportunity to further learn about magnetic positions over time, it seems likely that adults at least do develop or refine a magnetic map.

Indeed, the Lohmanns and others have shown that green sea turtles (*Chelonia mydas*) use a magnetic map to navigate towards specific targets.[5]

Questions remain about the extent to which turtle navigation abilities develop with age and experience, and how flexible their ability can be. Not only can the Earth's magnetic field change over time, but by responding and modifying their magnetic sense with experience turtles could learn features of the environment as they age and move around. Relatively recent work on logger-head hatchlings has shown that individuals reared as embryos under natural magnetic fields orientate correctly, whereas those raised in nests under manipulated fields orientate at random when they hatch.[6] It is unclear how long such effects last, or why this actually happens, but it shows that the magnetic sense of turtles is influenced by magnetic information experienced during development.

Staying in oceanic currents is clearly very important for the turtles. But one of the incredible feats in turtle navigation is that mature females return to the exact same beaches where they hatched in order to themselves reproduce, many years later. Roger Brothers and Kenneth Lohmann were able to show that hatchlings imprint on natal beaches, and then adult females return to the same place by cuing into its specific magnetic signature.[7] They analysed a dataset spanning 19 years of nesting behaviour and magnetic field information corresponding to different beaches. This revealed that as the Earth's magnetic field subtly drifts over time, there are corresponding changes in turtle nesting densities in beaches where the magnetic profiles of neighbouring beaches converge or diverge. That is, turtles seem to move around areas in their nesting patterns in line with small changes in the local magnetic fields that occur while females are at sea. This change in nesting location is predicted if turtles imprint on the specific magnetic profiles of the beaches where they hatched. Furthermore, some beaches, even those that are geographically separated, will have similar magnetic profiles, sometimes even more comparable to one another than to beaches that are closer in space. This means that if turtles imprint on natal beaches using magnetic information, females may sometimes end up making a mistake and nesting on the 'wrong' beach on their way to their true target. Consistent with this, Brothers and Lohmann found that the genetic structure of populations of turtles across a wide range of locations was better explained by the magnetic similarity of beach profiles, rather than their actual geographical separation; turtles that nested on beaches with similar magnetic signatures were more closely related, even if those beaches were not geographically close to each other. This was a clever piece of research because it used a long-term dataset and population analysis to show how females find their way back to nesting sites. A more direct way to do this would be to rear turtles on one beach but under magnetic information from another site, and see to which beach they return to breed. Clearly, given the length of time it takes for females to reach a reproductive age, among other logistical problems, such experiments are far from trivial to do. The study above may be less direct, but nonetheless it helps answer the question very effectively.

Magnetic information, combined with other senses, allows loggerhead turtles to perform remarkable feats of navigation, enabling individuals to find and remain in warm currents and feeding grounds, and then years later to head back to their natal beaches to breed. Loggerheads are by no means unique in doing this, and many other turtles follow long-distance migrations. Leatherback turtles (*Dermochelys coriacea*), the biggest of all turtle species, being up to and above 2 m in length, have been recorded to travel thousands of kilometres. Satellite tagging to track the movement of twenty-five leatherbacks showed that they swam a loop of 6,000–12,000 km from the temperate waters of Canada/ northern USA to tropical areas off South America and the Caribbean, and back again.[8] Such long-distance migration patterns in a variety of animals is likely driven by the need to find key feeding grounds, warmer waters, and places to reproduce.

That turtles and a vast range of other animals use a magnetic sense is not debated. What is contested, however, and in fact forms one of the most troublesome problems in all of biology, is just *how* the magnetic sense actually works. To delve into this, we need to move on to two other groups of animals, many species of which also undertake remarkable migrations: fish and birds.

It is no exaggeration to say that understanding how the magnetic sense of animals actually works has been one of the great mysteries in biology. A major problem lies in identifying what the magnetic receptors are. In most senses, the broad anatomy and location of the sensory system is relatively obvious, since it must to be open to the external environment to let information in. In vision, light has to pass through openings in the eyes to the photoreceptors; in hearing, sound has to pass through openings to the inner ears; in olfaction, the molecules must physically contact the cells that detect different chemicals; and so on. Generally, the organs responsible for these senses are clear. In contrast, because everything is immersed in the Earth's magnetic field, including the bodies of animals, in theory magnetic receptors could be anywhere—the head, the legs, or even the intestines! This makes it difficult to know where they are, let alone how they work. A further challenge is that the magnetic sense tends to be used in orientation behaviours, and here animals also make use of other senses. For example, in navigating long distances, birds might also rely on starlight patterns, olfaction, and visual landmarks. Designing experiments to test and isolate the role of magnetic cues is not trivial, not least if animals switch between these senses. In fact, it can be one of the reasons why similar experiments sometimes arrive at different findings. Nevertheless, in the past 15 years or so, major progress has been made, and there are currently two leading ideas for how magnetoreception works in most species that both have support from evidence. These two theories relate to either the presence of crystals in the body that react to magnetic fields or to light-dependent chemical processes occurring in the eyes in response to magnetic information.

The first theory involves a mineral with a name that would be well suited to a Superman film: magnetite. This is a type of iron oxide (Fe_3O_4), which forms a permanent magnet, and naturally magnetized pieces have long been known as lodestone. Its magnetic properties arise from the arrangement of the crystal lattice, and of the grains; magnetite grains in molten rock can align themselves along the Earth's magnetic field lines, 'freezing' the field into the rock as it cools. We have known since the 1970s that magnetite occurs in bacteria and other unicellular organisms and plays a role in how they respond to magnetic field lines. Around the same time, scientific studies, especially work on birds, sharks and rays, and honeybees, started to show more definitively that animals could respond to the Earth's magnetic field. By the early 1980s magnetite had been found in a range of organisms, not just bacteria, but also animals including honeybees, pigeons, and dolphins. This raised the possibility that it could form a sensory system, in which changes in the arrangements of magnetite grains responding to the external magnetic field could trigger receptors similar to those that are used to detect movement or changes on pressure. Or, the magnetite grains could even be positioned within cell membranes, such that they move position and open up membrane ion channels, triggering nerve impulses directly. In many ways this theory makes sense since, as we will discuss shortly, in birds magnetite has been found in deposits around the nose that are associated with the trigeminal nerve, and in other animals such as crocodiles and alligators, this nerve processes information from receptors detecting pressure and touch. The magnetite mechanism is often implicated in the use of magnetic maps.

The second theory for how a magnetic sense might work is normally called the radical pair theory. There's no getting away from the fact that this one is rather complex, involving a mixture of physiology, chemistry, and quantum physics. We need not worry about the detail here, but in summary the theory goes as follows. Light excites certain chemicals in the eye (we will come to which ones later), leading to a process whereby 'radical pairs' are formed. (A radical pair consists of two radicals—molecules with an unpaired electron—which are quantum entangled.) A radical pair can exist in different quantum states depending on its magnetic alignment, so the states of the radical pairs formed by light excitation are affected by the orientation of the Earth's magnetic field. The photoreceptors in the eye respond both to the presence of light, and the states of the radical pairs, and these latter potentially modulate the receptors' normal responses to light. Owing to the fact that the retina of the eye is curved, each receptor faces a different direction, leading to different inter-actions with the magnetic field and patterns of activation in space. One possible outcome of this is that the animal could somehow remarkably visualize magnetic information. The radical pair process could enable individuals to detect the inclination of the magnetic field and use it, for example, in a magnetic compass. One of the key predictions of this theory is that the magnetic sense is dependent on the presence of light. We can fill in some of the details of these two theories as we explore this topic more, but for now this brief summary should enable us to move onto our next two animal groups, and the evidence that they provide for each theory.

It is not only turtles that achieve remarkable feats of navigation. A number of fish species also travel great distances during different phases of their lives, often returning to natal spawning grounds to breed later on. One such fish is the Pacific sockeye salmon (*Oncorhynchus nerka*). In the spring, eggs hatch from their gravel river beds and the fry use odour cues to navigate their way through river systems to reach key nursery lakes where they grow for the first year of life. They then migrate to the Pacific Ocean, sometimes thousands of kilometres away, and remain there for up to several years before returning to natal spawning grounds, their original riverbed, where after spawning they die.

In the first stage of their life, the fish fry must move from their stream to a suitable area of a lake, and to do this they seem to use some sort of compass to travel up through the lake to a desired area. Thomas Quinn at the University of Washington established in the early 1980s that the fish can do this at least partly using magnetic information.[9] When placed in a tank with four arms, fish fry orientated themselves in a direction consistent with where they would be swimming naturally on reaching their lake. However, when a magnetic coil was used to redirect the magnetic field by 90 degrees, the fish changed their preferred direction broadly in line with this. They did not show these same changes when they could view the sky during the day but only when the sky was obscured, suggesting that visual cues of the diurnal sky take precedence over magnetic ones. In contrast, celestial cues seemed to be less important than magnetic ones at night because the fish responded more to the magnetic information regardless of whether the sky was visible or not. The magnetic sense in these fish could be most valuable for the fry when they have to move under snow and ice early in the year when they cannot see the sky, as well as in those that move at night. So, the salmon use magnetic information to guide their lake movements, but Quinn also suggested that, logically, it could be valuable for guiding oceanic migrations too.

Scientists have now investigated how salmon find their way back from the ocean to the area of their home rivers. We know that in many cases their final navigation upstream to natal spawning sites seems to be reliant on smell, but how they get to the start of their rivers on the coast has been less clear. Sockeye salmon apparently do this by imprinting on the local magnetic field where they enter the ocean, and then find their way back to this position later on. Scientists have analysed over 50 years of data from salmon fisheries, comprising measurements of the proportion of salmon that pass around Vancouver Island, travelling either to the north or south to get to their natal Fraser River, combined with data on shifts in the local magnetic field, rather like the study of turtle nesting patterns.[10] The route the salmon took could be predicted by drifts in the Earth's magnetic field between the two passages and the mouth of the river; as the magnetic field shifts, so does the primary route the fish take. The scientists speculated that the salmon may use a magnetic map to encode the location of the river entrance based on magnetic field intensity and inclination. Studies of juvenile Chinook salmon (Oncorhynchus tshawytscha) show that they too respond to simulated magnetic fields consistent with different ocean latitudes—these would enable them to find feeding grounds. Like the turtles, these fish had no previous experience of migration, so their magnetic ability appears to be inherited.

It's clear that various species of salmon can respond to and use magnetic information, so how do they do this? In 1988, electron microscopy work on sockeye salmon revealed the presence of magnetite crystals in the head of the fish. The structure of the crystals and their arrangement was consistent with their being secreted by the animals for a purpose, rather than occurring simply as an artefact of other processes.[11] The magnetite seems to be produced during the lives of the fish, and is present in sufficient quantity to allow the detection of small differences in the intensity of the Earth's magnetic field. Not much work has been done since on this species, but other fish have been well studied for their magnetic ability, including the rainbow trout (*Oncorhynchus mykiss*).[12] These have also been shown to perceive and respond to magnetic fields in orientation behaviour, and scientists have identified possible magnetic receptor cells in the nose. The cells, containing crystals shown to be magnetite, are associated with the trigeminal nerve and respond to the intensity of the magnetic field. Moreover, other studies using electrocardiogram measurements of heartbeats have demonstrated that fish perceive changes in the magnetic field, and that damaging the branch of the trigeminal nerve, which leads to the nose, affected the fishes' ability to do so.[13] Their responses were also independent of light conditions. At this point, the evidence that fish such as salmon and trout rely on a magnetite-based system was convincing. While it had yet to be shown how the magnetic crystals triggered nerve responses in putative magnetic receptors, much of the evidence was there.

One of the main methods that has been used to investigate a magnetite mechanism, especially in birds, is to apply a short magnetic pulse to the location where the magnetite particles seem to be. This temporarily changes the structure and arrangement of the particles and should prevent the magnetic system from working properly. On the other hand, a light-dependent magnetic system should not be affected. Combined with this approach, other studies have focused on understanding the genes involved with trout magnetoreception. The application of a magnetic pulse leads to changes in the expression of a wide range of genes. Knowing which genes are affected can help to reveal the basis of the magnetic sense because we expect genes linked with certain magnetic processes or tissues to be affected in their activity when subjected by a pulse (or indeed longer-term stimulus). Some of the changes found in gene expression do seem linked to processes involved with the repair of magnetite-based receptors. By contrast, analysis of the impact of a magnetic pulse on the expression of genes from tissue from the retina in the eyes has found very little effect on gene expression at all.[14] So there appears to be little evidence for magnetite-based receptors in the retina. An outcome like this is what we would expect given that the candidate magnetite receptors have been found in the nose of trout (and perhaps the brain too). That said, it may be premature to rule out a light-dependent magnetic sense occurring in the eye of fish like trout, since magnetic pulse experiments would not be expected to affect the light-dependent mechanism. Nonetheless, there is very little evidence for this, and overall the evidence is consistent and quite convincing in showing a magnetite-based process as either the only or main mechanism used in fish. So far so good, but this is not where the story ends.

Suggestions that birds might use a magnetic sense go back to the 1800s, yet it took over 100 years for clear evidence to arise. Considerable effort and much important work on magnetic-based orientation in birds has been undertaken by Roswitha and Wolfgang Wiltschko at Goethe University, Frankfurt. In 1968, Wolfgang Wiltschko showed that European robins (*Erithacus rubecula*), a species that migrates at night, use a magnetic sense to determine direction, because their preferred direction could be altered by changing the magnetic field.[15] Soon after, in 1972 the pair demonstrated that robins measure inclination. The Wiltschkos utilized a type of behaviour many migratory birds show at certain times of year called 'migratory restlessness'. When the birds would normally be migrating, those kept in cages tend to show behaviours that orientate towards the direction they would naturally intend to travel (Figure 30). For example, if during the spring the birds normally travel north, then those in cages might jump and scrabble towards the north side of their cage. The degree of restlessness and the predominant direction in which the birds try to fly can be measured, by recording how often birds jump on different perches or by analysing the scratches made by individuals scrabbling around on typewriter correction paper. In robins, the birds' behaviour did not depend on changes in the polarity of the magnetic field, but rather on its inclination, with an internal inclination compass telling the birds whether they are heading towards or away from the pole or equator.

Shortly after, research on garden warblers (*Sylvia borin*) showed that individuals raised without seeing the sun or stars could still orientate appropriately, revealing that they too use a magnetic compass.[16] In reality, many birds utilize other senses too, including starlight, Sun, and olfactory information, and they will cross-calibrate these systems to allow for greater accuracy on different points along their movement pathways. However, the work with garden warblers showed that the magnetic sense could work and provide key information at least somewhat independently from these other senses.

Over the next 20 years or so, many further studies showed that a variety of avian species, particularly but not limited to nocturnal migrants, use a magnetic sense. They range from pied flycatchers

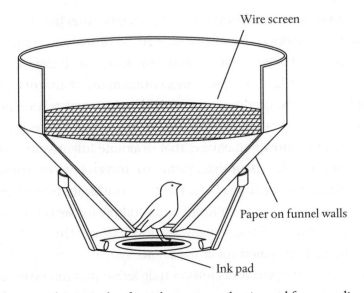

Figure 30. Classic Emlen funnel apparatus that is used for recording avian migratory restlessness and measuring the main direction that birds show propensities to travel during migratory periods.

(*Ficedula hypoleuca*) to savannah sparrows (*Passerculus sandwichensis*). It is tempting to think of the magnetic sense as something only used in long-distance behaviours, but not all of the birds that use a magnetic sense are migrants; pigeons for example use their magnetic sense for homing behaviour in returning to their roosts. The avian magnetic sense is now widely appreciated to play a role not just in long-distance movements, but also over much shorter spatial scales, including home range behaviour. All in all, over twenty bird species have been clearly demonstrated to use magnetic information as a compass and to respond to different components of the magnetic field. The actual number is likely to be far greater.

The question that concerns us here, though, is how this avian magnetic sense works. Early studies of pigeons in the 1970s, involving placing magnets on the back of the birds' heads, caused disorientation under cloudy skies, as would be expected under some sort of magnetite process.[17] Work soon after tested a long-distance migrant, the bobolink (*Dolichonyx oryzivorus*), also showing that the birds responded to changes in the magnetic field.[18] On examining the bodies of the birds, researchers found magnetite deposits around key nerve systems and in bristles that projected into the nasal cavity, as well as small deposits in parts of the brain. The deposits seemed to be in close proximity with nerve cells. Magnetite has since been further located in the inner skin around the upper beak, and structures that contain magnetite and similar substances, again seemingly linked to nerve cells, have been found in a range of avian species.

The next line of evidence for a magnetite system came from experiments applying a magnetic pulse to birds' heads. As we know, this should affect the properties of the magnetite particles, temporarily affecting the magnetic sense. Indeed, applying a magnetic pulse to the heads of silvereyes (*Zosterops lateralis*) resulted in a 90-degree shift in orientation behaviour, which returned to normal after about one week.[19] Similar effects in disrupting orientation have been found in studies on a variety of birds from migrating birds to homing pigeons. These findings point to the use by birds of a magnetic map that relies on a magnetite system. Interestingly, disruption experiments often produce effects that tend to be found in older but not juvenile birds, suggesting that the system involves some aspect of experience during the birds' lives. This again is consistent with a magnetic map, which although it can have an innate component will often involve some aspect of learning or imprinting on local magnetic information over time.

At this point, the logical conclusion would seem to be that the key evidence for how the avian magnetic sense works is based on a magnetite process. Alas, the magnetic sense has never been straightforward to understand, and alongside magnetite, research studies testing the alternative light-dependent theory were also gathering pace. In fact, it was already known that migrating birds often perform what is known as 'head scanning behaviour' when using the magnetic field. The birds make characteristic head movements, looking around them during this time. It is almost as if they are trying to 'see' magnetic information. In the mid-1980s, more experiments on pigeons showed that parts of the brain associated with the visual system of the birds responded to the presence of magnetic information, specifically the direction and intensity of the magnetic field. Initially, this seems very odd—parts of the brain used for vision are being stimulated by magnetic information.

As it happens, a potential magnetic system that is dependent on light had already been suggested to facilitate magnetoreception, and the first clear evidence in birds came in 1993 in a study on silvereyes led by Wolfgang Wiltschko.[20] This bird migrates between the Australian mainland and Tasmania, especially during dusk and dawn. Birds were tested for their ability to orientate under magnetic cues when in the presence of white light, or under light that was restricted to red, green, or blue wavelengths only. Under all conditions except red light, the birds orientated in a north-northeast direction, whereas under the longwave red light, there was no consistent directional pattern. At the time it was not well known to the authors how effectively birds see different wavelengths of light, but we know clearly today that birds can see all the wavelengths presented in the study. Therefore, the lack of orientation behaviour under red light was not simply due to the birds being unable to see the light, and effectively being in darkness. Instead, magnetic orientation behaviour was strongly influenced not just by the presence of light, but by the actual wavelengths of light present.

A large variety of follow up work has confirmed and extended these findings. Two years later Roswitha and Wolfgang Wiltschko showed that European robins also orientated appropriately to magnetic information under white and green light, but not under red light (Figure 31). Intriguingly, when they also applied a magnetic pulse to the birds, many but not all of them were affected by this but in different ways, including changing direction. The study therefore was consistent with birds having *both* a light-dependent compass and a magnetite-type map. Likewise, further work revealed that while robins could orientate to the magnetic field under 565 nm green light, they did not show directional preferences under 590 nm yellow light; there was a narrow region of

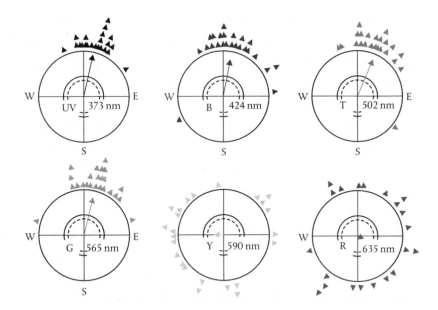

Figure 31. Experiments on the orientation behaviour of European robins in spring under light of different wavelengths. Under shorter wavelength blue, turquoise, and green light, the birds orientate consistently for their migratory direction, whereas under yellow or red light their orientation behaviour is abolished. Triangles represent the mean headings of individual birds, and arrows represent overall average direction across individuals.

light wavelengths where this switch in light dependency occurs. Later work even narrowed down an important region between 560 nm and 567 nm between which robins change from showing appropriate orientation behaviour to not doing so. But the situation is not straightforward, since the responses also vary depending on the intensity of the light.

It is clear that a magnetic compass in many birds seems to need the presence of light, specifically short to medium wavelengths. This does not tell us though how the sense actually works. Over the years, scientists have suggested a variety of different ideas, but gradually they have honed in on the likely candidate: a group of molecules with another suitably science fiction sounding name: 'cryptochromes'. Cryptochromes are protein molecules found widely in nature and early on they were suggested as possible candidate molecules underpinning the light-dependent magnetic sense; they form radical pairs when activated by light and are sensitive to shorter UV or blue wavelengths.

The evidence for the importance of cryptochrome gathered pace in the mid-2000s onwards when a spate of studies discovered a number of smoking guns that were all consistent with cryptochrome being the key proteins. In 2004, Henrik Mouritsen and colleagues from the University of Oldenburg in Germany showed that garden warblers (S. borin) have types of cryptochrome that occur in the eyes. In particular, cryptochromes were concentrated in certain types of cell that play a role in processing early-stage visual information in the retina—the ganglion cells.[21] Furthermore, these cells were strongly activated at night, something that is consistent with when the warblers show orientation behaviour in response to magnetic information. By contrast, non-migratory birds (zebra finches; Taeniopygia guttata) did not show the same levels of activity associated with cryptochrome, and it was largely absent from the eyes. Therefore, although the study did not show directly that cryptochrome is used in magnetic orientation, several lines of evidence were consistent with this. In 2011, a particular type of cryptochrome was also found in the ultraviolet-sensitive cone cells of birds that are known to use a magnetic sense (robins and chickens), distributed across the eye.[22] The molecules were also arranged along cell membranes in such a way that should make them functional in fulfilling a role in the radical pair model. Very recent work has also discovered evidence for the use of other types of cryptochrome molecules in migratory birds, found in other cone types, in that the molecules occur at higher levels in the eyes of migratory birds at times of year when they would be migrating. Different forms of cryptochrome molecule may be used in the magnetic sense, and work is still progressing in this area, including importantly how they actually enable the sense to work. Yet regardless of these missing details, cryptochrome itself is currently the only group of protein molecules that fit the requirements of the radical pair model and much research supports their use in this.

I mentioned the evidence as being in the form of 'smoking guns' because work has as yet to definitively show the actual receptors involved and how they are triggered. Nonetheless, research by the same team at the University of Oldenburg and others has also shown that a part of the avian forebrain seems to process information from the light-dependent magnetic sense: an area called 'cluster N', and this area is part of the visual system and seems to be active in migratory but not non-migratory birds.[23] Experiments have also shown that 'cluster N' is connected to the nerve cells in the eyes of the birds and is active during magnetic orientation behaviour. In robins, lesions to cluster N prevent the birds from showing magnetic orientation, whereas damaging the trigeminal nerve did not have the same effect. So, while we might not yet know the details of how the initial stages of the light-dependent magnetic sense works, we do know that a part of the brain that plays a role in vision seems crucial in enabling magnetic compass orientation.

The early 2000s and beyond marked an intense period of research into the basis of the magnetic sense in birds. As time went on, support for *both* the magnetite and light-dependent theories gathered pace. For example, studies of robins indicate that they seem to use different magnetic mechanisms and respond to different features of the Earth's magnetic field under certain wavelengths of light versus when in darkness. Other studies show that the avian magnetic compass, responding to inclination, is not based on a magnetite system. Instead, magnetite seems to provide information most often on magnetic field intensity. In birds, the magnetite system apparently underpins a magnetic map based on intensity, whereas the light-dependent system facilitates a magnetic compass based on inclination. However, we will have to wait a bit longer for direct proof of the actual pathways involved and the specific receptors used. Just to make things more complicated, some scientists have asked whether some of the structures found in the nose, thought to be used in a magnetite-based magnetic sense, are actually just by-products of the immune system, and other studies have suggested a role of sensory structures found elsewhere in the head. It has even been suggested that magnetoreceptive symbiotic bacteria living inside the tissues of animals like birds may provide them with their magnetic abilities. It is an intriguing idea, but highly speculative and does not really solve the problem of what receptors birds actually use. One thing is for sure: the mystery of how the magnetic sense works is not solved quite yet.

Over the years, a magnetic sense has been demonstrated or suggested in numerous other animal groups, from mice, bats, mole rats, glass eels, ants, moths, cockroaches, sharks, fruit flies, lobsters, and beyond. Such discoveries support both the use of magnetic compass and map information, and of both magnetite and light-dependent systems. For our focal animals in this chapter, in fish and sea turtles the evidence is that they rely on magnetite substances or something similar. In birds, the weight of evidence, despite some recent challenges and complications, indicates that they use both a magnetite map and a light-dependent compass, with the strongest evidence for the latter. None of these stories are complete yet though. Resolving the remaining issues will require some effort. To quote one recent scientific article concerning understanding how magnetoreception works: 'solving this scientific mystery will require the development of new genetic tools in magnetoreceptive species, coupled with an interdisciplinary approach that bridges physics, behavior, anatomy, physiology, molecular biology, and genetics'.[24] That is no small task!

It is hard for us to get a handle on the magnetic sense because it is so far from our own perceptions. Not only is it widely used and often very refined, but the way that it works is potentially complex and diverse. As far as sensory systems go, it also throws up some rather weird findings. For example, scientists have analysed satellite images of cows and deer, and shown that herds align themselves with magnetic information.[25] Specifically, animals in herds seem to orientate on a north-south axis with regards to magnetic north. Quite why, nobody knows—it might help the animals to keep track of directions or to all run in the same direction if a predator approaches, rather than crashing into one another. Or perhaps there is no real reason at all, and not every study has been able to replicate those findings. Certainly though, a variety of animals show magnetic alignment, and it may even play a role in things like the hunting behaviour of foxes, possibly enabling them to increase precision when jumping on prey. Even more bizarre, dogs have been shown to urinate along a north-south axis, seemingly based on magnetic cues, specifically polarity.[26] Quite why they do this, who knows? Let alone what led the researchers to analyse 5,582 observations of the direction in which seventy different dogs urinated. Many mysteries remain. On that note, we come to the end of our tour of the main senses. In Chapter 8, we deal with probably the biggest contemporary issue of all: how human impacts on the world are affecting animal senses, and what we can learn from them for our own societies and inventions.

CHAPTER

8

SENSING IN THE ANTHROPOCENE

Our growing understanding of the variety of sensory systems used by nature has inspired new technologies to benefit our lives and society, and we will look at one or two of these in this chapter. But even as we improve our lifestyles we must be aware of the impact we produce. The living world of which we are a part is changing at an unprecedented rate, and the reason for that is humanity itself. There is probably no habitat on Earth that is not being affected by humans in one way or another—through general destruction of the natural environment, climate change, or a plethora of pollutants. Our reach is planet-wide. From plastic found even at the very bottom of the oceans, to large amounts of noise and light occurring in virtually every place we inhabit, we are affecting organisms in a multitude of ways. One of the major consequences, and one that we are increasingly coming to appreciate, is that we are dramatically altering the sensory worlds of animals. In recent times an increasing number of scientists have scrambled to try to understand this, and while solutions are tricky, we are at least beginning to recognize the damage done. Unfortunately, the solutions are often highly political, and require efforts well beyond the science itself. On a more positive note, we are also using knowledge of animal senses to help solve some of the problems we have created.

The idea that animal senses can play an important role in aiding humans is by no means new. Anyone who drives in the UK will know all about 'cat's eyes', the reflective structures embedded in the sides and centre of roads which allow us to see its boundaries at night. The design was inspired by the highly reflective structures (a *tapetum lucidum*) in the eyes of many nocturnal animals, including those of cats, which shine light back that would otherwise have missed the photoreceptors, enhancing visual sensitivity. And we have already noted that humans utilize impressive canine olfaction, with sniffer dogs employed for all manner of tasks, from finding drugs and hidden explosives to even trying to sniff out potential diseases in humans. Beyond these examples, there are many more ingenious ways by which humans make use of animal senses to solve problems, albeit including many created by us in the first place, and to advance technology.

War and conflict never seem far away from humanity, and one of the most terrible associated actions is the laying of landmines and other explosive devices. Aside from the obvious horrors these produce, a major issue with them is that once any conflict is over, finding and getting rid of them is extremely hard and risky. They are hidden and often deployed over large areas, unmarked, and haphazardly placed. Since they can kill and maim any who inadvertently walk over them, the whole area must be avoided, and minefields can perpetuate the displacement of populations for decades after. Unfortunately, because mines are far easier to lay down than to find and destroy, the number of landmines globally is increasing rather than declining.

One solution may arise from the fact that landmines and other devices have characteristic odours that can be detected. The smell is too subtle for humans, and indeed for most technology, yet is well within the scope of a range of other animals and their vastly superior sense of smell. Over time, various animals have been tested and used to locate the presence of mines. One of the most widely used has unsurprisingly been dogs, with a certain amount of success in some parts of the world. However, dogs need a good deal of training, tend to tire quickly in the task of finding mines, and can be overwhelmed by multiple odour sources, which is a problem when there are multiple mines around.

Another, more unexpected, animal has begun to be used for detecting mines: African giant pouched rats (*Cricetomys gambianus*).[1] These rodents are already pre-adapted to many of the demanding climates where minefields are often found, they are more easily trained for repetitive tasks, and they have a better sense of smell than dogs. The rats are also less dependent on a specific trainer and easier to move around, and a major plus is that since they weigh only 1–2 kg, they rarely detonate the mines if they happen to walk over one. The rats are trained to find mines with a food reward such as banana, and kept on a leash in the field while searching. When they find one they pause, and paw or bite the ground. They have been used for mine detection in countries such as Mozambique; in 2009 a team of thirty-four rats was used to clear an area of nearly 200,000 m² and found seventy-five landmines, allowing 750 families to return home. Related studies of the performance of rats in the field have shown them to find 100 per cent of the mines in areas as large as 93,400 m². So, while it is early days, and the rats are not a magic solution to rapidly clear all mines, they do offer another valuable route to dealing with the problem.

An understanding of animal senses has also inspired new inventions. For instance, many deterrents for supposed pests are based on the animal's senses; controlling the risk of bites and disease transmitted via mosquitoes has long been aided by chemicals that target the mosquito's sense of smell. A growing area is biomimicry, which involves inventions inspired by nature, and this has included trying to copy the sensory systems of animals.

Our discussion of the tactile senses of animals highlighted the remarkable ability of crocodiles and alligators to detect changes in movement and pressure of water to find prey. This ability is conferred by the dome-shaped structures distributed over the head and body of the animals, which are highly sensitive to tiny changes in pressure and distortion. Engineers are now trying to create artificial sensors for detecting water movements based on their natural counterparts in crocodiles. The idea is very similar: a dome structure, 2 mm by 2 mm, houses electric sensors that respond to changes in the mechanics of the dome and to pressure in general from sensors on the dome surface.[2] These sensors can detect the direction and types of water flow encountered, even as far as the speed and direction of water movement. Other studies have focused on mimicking the ways that fish lateral lines sense water flow. Such sensors could be used on 'autonomous underwater vehicles' or AUVs, which play an important role in underwater exploration and in the military. The problem is that AUVs need a range of sensory systems to automatically respond to the environment. Visual cues are often no good because of poor water clarity, and sound devices like sonar can be obstructed by blind zones, and potentially harmful to aquatic life (or, indeed, they risk detection). What's needed are sensors that use minimal energy (to keep the AUVs light) yet give accurate pictures of the surroundings. Work is ongoing, but given how acute the sense of crocodiles and fish is in detecting water movements, any device that can mimic it is likely to be of great value.

Another example of biomimicry involves attempts to replicate the abilities of animals like cuttlefish to change colour rapidly in response to visual information. Cuttlefish, squid, and octopus are fabled for their prowess in changing their body patterns within seconds in order to resemble their background and be hidden from predators. They do this by having special cells called chromatophores, within which are pigments that can be contracted or dispersed to change the colour of the animal. Many species of animal can change colour, but most do this fairly slowly (like shore crabs). In cephalopod molluscs such as cuttlefish, however, the chromatophore cells are under direct control of the nerve cells, which in turn are influenced very rapidly by what the animal sees, meaning that changes in body appearance are extremely fast. For humans, the main potential benefit of copying this is in being able to create devices that automatically and rapidly respond to features of the background and lighting conditions—imagine tablets and mobile phones and other visual displays in which each part of the display adaptively responds to conditions, or the science fiction sounding world of artificial skin that could be used as cloaking devices, or even dynamic clothing for fashion. Recent innovations have involved creating artificial chromatophore cells that respond to the surroundings and change accordingly, and which can be grouped into arrays of cells.[3] In truth, at present these are fairly basic, and not exactly up to cloaking device levels, but the potential for materials to automatically and rapidly adapt to the environment offers striking opportunities as technology develops.

Given the incredible abilities that many animals have to sense the world, there are surely countless other innovations that would stem from trying to mimic those too.

While we may be inspired by nature in creating technologies to improve our lives, we also use animals directly in many ways. Tens of billions of birds are kept for food production, not to mention other animal groups. Understanding their sensory world can, and should, help us to give these animals conditions that enable them to stay in good health, and this is illustrated by work on lighting.

We humans are used to living and working with a variety of lights. Many of these lights do not emit continuously, but actually flicker on and off relatively quickly. This has long been the case for widely used fluorescent lighting. Normally, because the rates of flicker are quicker than our visual system can detect, we do not notice this. We can think of our visual system as having a limit in how quickly it can respond to changes over time, a bit like a refresh rate on a computer screen; this is called the flicker-fusion frequency. In most humans, the critical value is somewhere around 50–60 Hz, though it is influenced by various factors. As long as the lights we are using flicker on and off at a faster rate, we would see this as continuous illumination. When that does not happen (such as with a faulty or old light) we can see it flicker, and this can be very unpleasant. In fact, flickering of lights and displays, such as that of computer screens, is linked to bad moods, headaches and migraines, and even epileptic fits.

The lights we use have naturally been designed with human vision in mind, but the flicker-fusion frequencies of other animals varies. In birds it may be as high as and beyond 100 Hz (though estimates vary), and this potentially creates a problem, since many standard office lights flicker at 100 or 120 Hz. To the birds it may be a little like living under a strobe light. Research at the University of Bristol in the 2000s investigated the influence of light flicker on the well-being of birds such as European starlings (*Sturnus vulgaris*).[4] An initial study by Verity Greenwood and others tested the preferences of starlings for lights that flickered at frequencies of 100 Hz versus very high-frequency lights that flickered well beyond the range of any animal (30,000 Hz). Although they found no evidence from hormone analysis that the birds were stressed under the low-frequency lights, starlings discriminated between the lights and preferred to reside under the high-frequency type. The research team also found evidence that birds being used in other studies, which were investigating their vision and behaviour more generally, seemed to sometimes incur some sort of involuntary muscle spasms when seeing visual (CRT) displays and low-frequency lighting. Such displays include older TV and computer monitors, meaning that even pets at home may be affected. Further work by Jennifer Evans and others tested how mate choice in starlings might be affected by lighting. They found that while the birds displayed normal and consistent mate choice behaviour and preferences for features of male plumage under high-frequency lights, this was negatively affected under low-frequency lights. Moreover, under low-frequency lights the birds showed higher levels of muscle spasms, and reduced jumping, eating, and drinking behaviour. And the birds were found to show changes in their levels of stress hormones, indicating chronic stress under low-frequency lights.

Not all birds will perceive the same levels of flicker. Some may not be so adversely affected by low-frequency lights, but the work above clearly shows that it can have significant negative effects. And further studies demonstrated that bird behaviour is also affected by a lack of UV lighting. So, if we want to keep animals like birds under comfortable conditions, one of the requirements should be for full-spectrum, high-frequency lighting. There is certainly increasing attention given to the housing conditions of animals with regards to their sensory capabilities, and many birds are now kept under higher-frequency lighting systems. As it happens, this does not seem to have come directly from a consideration for animal welfare, but rather costs. LED lights, for example, which are brighter and do not flicker like standard fluorescent ones, are much cheaper to produce and replace. So, changes have happened, if not for the reasons we might always expect. Regardless, there is much more potential for food industries and places such as zoos to adopt conditions that are tailored to the sensory worlds and well-being of the animals they keep.

The issue of flickering lights is not just relevant to captive animals. Much of the lighting used outside also often flickers on and off. Scientists at the University of Exeter collated a large database of published research on animal flicker-fusion frequencies and showed that many species may be able to perceive the flicker of a variety of sources, and this may sometimes include those used in night-time lighting.[5] Not only is this potentially harmful to animals in the ways discussed for starlings, but it might also explain why some animals are less frequently found and move less under artificial lighting. That said, the move to LED lights may to an extent solve this problem, albeit while creating others. We will consider the impacts of lighting on the natural world more generally later, but before we leave the subject of captive animals, let us consider how knowledge of animal vision can also help potentially improve animal welfare in more unexpected areas.

In my own research, our lab at the University of Exeter studies animal vision and behaviour. We have for some time been extending this knowledge and developing imaging technology that can analyse the world as it appears to other animals and which can be applied in various areas. One example has been to test how horses see the colours of fences and hurdles in horse racing, in work undertaken primarily by Sarah Paul. Horseracing remains a popular event around the world, and in the UK and many other countries a common format involves jump racing, where the jockey and horse have to jump over fences or hurdles in racing round the course. The sport is not without its critics, not least because horses sometimes get hurt when falling at obstacles. In the UK, a long tradition dictates that nearly all fences have visibility markers (take-off and midrail boards) that are painted orange, set against the green-brown vegetation of the rest of the fence. The problem is that horses are classic dichromatic mammals—they cannot see the difference in colour between green and orange that would be apparent to most jockeys and trainers. For our research, we visited race courses around the UK and took special images of the fences that could determine how well the horses would see these visibility markers, versus the use of other possible colours (Plate 9).[6] We also tested in a training environment whether horses jump differently when faced with markers of other colours and contrasts. As expected, the orange colours were not easy for the horses to see, and other colours should be much more visible, especially white, some blues, and fluorescent yellows. In the behavioural trials, horses changed the angle and distance at which they jumped fences when they were marked with blue, yellow, or white compared to orange. This work, which was funded by the racing industry to address this shortcoming and ideally improve safety, is still on-going. The governing body of UK racing (the British Racing Authority) is now conducting larger trials with a view to potentially making changes to actual race courses. Whether that happens, and what effect it has on actual casualties only time will tell, but there can be little doubt that we can improve the lives of captive and sport animals by understanding how they perceive the natural and human environment. And if looking at the world from the point of view of sports and captive animals helps us to recognize more fully how our activities impact on them, our knowledge of animal senses can be used more widely, to reveal often hidden effects of our lifestyles on the natural world.

Of all our impacts on the natural world, besides general habitat destruction, chemical pollution is probably the area that has received the most long-standing attention. Humans have known for decades that chemical pollution is a major problem for aquatic life. Chemicals entering the waterways from processes such as industrial plants and agricultural run-off upset the eco-system balance in a multitude of ways. Pollutants are often directly harmful or toxic to life, or they cause some organisms to thrive on the altered nutrient balance in the water at the expense of others. A great deal of research has focused on these issues, not least in testing the concentrations at which certain chemicals become lethal for aquatic life. But problems caused by pollution can be problematic in other, sometimes less expected ways too, including by interfering with animal senses and having knock-on effects for a variety of behaviours.

Over a period of time beginning in the early 2000s, scientists have investigated how crayfish behaviour and communication is affected by agricultural pesticide chemicals leaking into the water.[7] This work has shown that chemical pollution does not just cover up important chemicals normally used by crayfish in their sense of smell, but that it also affects the sensory receptors themselves. It can even produce maladaptive responses. An initial study by Mary Wolf and Paul Moore from Bowling Green State University in 2002 investigated how the behaviour of a species called the rusty crayfish (*Orconectes rusticus*) was affected by sub-lethal levels of the pesticide, metolachlor. This is a herbicide used to treat weeds around crops of corn, soybean, cotton, potatoes, and others, and has been implicated as causing harmful side effects in humans and other animals at higher levels. Crayfish are heavily reliant on smell and chemical communication in their daily lives. They are a large species, up to 10 cm in length, and often aggressive to one another. And behaviours such as establishing dominance hierarchies between individuals, finding food and mates, and alerting others to threats, are all communicated chemically (Figure 32). Wolf and Moore tested how crayfish responses to the alarm signals given off by other injured crayfish were affected by the presence of the pesticide. They found that individuals exposed to it showed highly mal-adaptive responses: they walked faster and *towards*, not away from, alarm signals and danger. In other words, the crayfish did not simply respond less to alarm signals, but actually reversed their behaviour and approached danger instead of avoiding it when in the presence of the pesticide. The crayfish were also less

Figure 32. Rusty crayfish (*O. rusticus*) compete with one another for dominance and in other social interactions. Here two adult males fight for dominance. Many of these are mediated by chemical signals released from another. However, pesticides released into the environment can interfere with their sense of smell, changing these interactions.

able to find a food source when the chemicals were present. Why these changes occurred was not understood, but it was likely due to the way in which the pesticide may interfere with how the sensory cells function.

Six years later, work by Michelle Cook and Paul Moore reported how metolachlor also affects aggressive interactions. Crayfish that had been exposed to the pesticide for 96 hours, and then allowed to interact with non-exposed individuals, were less likely to initiate fights and less likely to win contests than those that had not been exposed. Again, it is likely that the pesticide chemicals alter in some way the sensory receptors' ability to encode chemical signals from a rival. Further work from the same laboratory showed that crayfish are also badly affected by copper pollution, which can originate from sources such as coal power stations and metal production as well as from agricultural chemicals. That chemical pollution can affect communication and behaviour in a range of aquatic organisms is now widely established. Among swordtail fish (*Xiphophorus birchmanni*), named for their elongated tails used as a signal in mating, water that contains sewage and agricultural run-off can lead to fish failing to show mate preferences for their own species.[8] Beyond these and other cases, disruption of animals' sense of smell, and consequent interference with key behaviours, can also arise from more unexpected routes, such as marine plastic.

Birds have not always been known for their sense of smell. While their vision and magnetic sense, for instance, is revered, there has often been a misconception that birds have a poor olfactory ability. Certainly, smell is unlikely to be the primary sense in many bird species, but this viewpoint is at the very least an oversimplification. A number of bird groups are now recognized to have a remarkable sense of olfaction, which they use in finding food and in navigation, among other things. Some of the most gifted birds are the 'procellariiform' seabirds that include groups such as albatrosses, shearwaters, fulmars, and petrels. Many of these are known to be quite smelly themselves, and seem to use odour in mate choice and doing things such as finding their home burrows when returning after foraging. Preventing individuals from smelling can affect their nest finding behaviour and mate selection. These birds often travel great distances, sometimes thousands of kilometres, in search of food, which includes fish, crustaceans, squid, and other aquatic prey close to or under the surface. Quite how the birds find patches of food in the vast expanses of the ocean has been a puzzle. In 1995, Gabrielle Nevitt from the University of California, Davis, and others showed that part of their ability stemmed from locating a specific compound called dimethyl sulphide (DMS).[9] This substance is often produced by phytoplankton when they are under attack from zooplankton, in particular when the phytoplankton cells are crushed, releasing a chemical precursor that gets converted into DMS. The zooplankton includes animals such as krill, which in turn is one of the favoured foods of some seabirds.

Nevitt and her colleagues undertook experiments in sub-Antarctic waters near the island of South Georgia, where the birds are common during the Antarctic summer. They released different kinds of small 'oil slicks' onto the water, some of which contained DMS, whereas others had no smell, or had the smell of cod liver oil. Birds such as petrels were attracted to the DMS and cod liver oil slicks, but not to the control ones. Many of the birds were also influenced and seemed to detect the presence of DMS released as aerosols from the sides of the boat. Later work, spanning more than 20 years, by Nevitt and others has reinforced our understanding that birds do certainly detect and show physiological and behavioural responses to naturally occurring levels of DMS. This is very interesting in its own right, and illustrates some quite incredible sensory biology and ecology in these animals and their foraging behaviours. For our purposes here, though, the twist comes in a study in 2016 by Matthew Savoca and several colleagues including Nevitt.[10] They showed that when small pieces of marine plastic are kept in seawater, DMS is produced. This happens as a result of 'biofouling', when algae grow on the plastic and then die, releasing DMS. Algae growing on plastic pollution, therefore, can produce chemicals that are interpreted by the procellariform seabirds as a sign of food. In line with this, the team found that birds that are more responsive to DMS (or at least have nesting behaviours that would suggest this to be the case) also show higher levels of plastic ingestion. While not definitive evidence, the study indicates that plastic consumption by seabirds may come from behaviours and a sense of smell that would normally be used to find food.

Not everyone agrees with the specifics of this particular study, but while some of those criticisms may be valid, they likely do not change the main story here. More questionable, though, is the suggested solution: to increase the anti-fouling properties of plastics. As critics have pointed out, the anti-fouling properties of products have their own environmental impact that needs addressing, and this solution fails to tackle the main issue of massive over-use and poor reuse of plastic by humans in the first place.

The problem of plastic in the environment, especially the oceans, is shocking—some have suggested that more than 99 per cent of all seabirds will have ingested plastic by 2050, not to mention the countless other animals affected, from fish to turtles. Clearly something needs to be done, and urgently. And much of the solution has to involve dramatically reducing our dependence on plastic. Recent actions have been positive in starting to address the problem of single-use plastics in items such as drinking straws and takeaway coffee cups, yet in truth much of the plastic pollution in the marine environment comes from items such as discarded fishing nets.

There are other sources of pollution beyond chemicals which can affect animal senses, and one of the most widely studied for its impact on behaviour has been sound or noise pollution. A multitude of sources of noise originate from human actions, not least noise from traffic, building work, and aeroplanes. At sea, ship noise, sonar, and seismic mapping of the ocean floor are all widespread disturbances. Noise itself is not actually new to most animals, and occurs naturally in many habitats. For example, when echolocating, many bats have to overcome the noisy calls of insects by shifting their echolocation frequencies, otherwise the noise might affect bats' ability to navigate and find prey. Rather, what's new in the modern world is the intensity of noise, and the frequencies involved, which are often novel and potentially detrimental to a whole host of animals.

Much research on the effects of noise pollution has been undertaken on birds, in part because of their celebrated use of song in communication to attract mates and defend territories. Perhaps we are also driven to study this by our own love of bird song. A study in 2003 by Hans Slabbekoorn and Margriet Peet at Leiden University in the Netherlands found that great tits (*Parus major*) in the city of Leiden sang songs at higher frequencies when their territories were noisy.[11] This showed that the birds could change their song, either as a short-term direct response to noise at the time, or perhaps based on the environment in which they have grown up. Doing so helps avoid overlap with city noise. This study was important but limited, since it only comprised birds from a single area, and so it was not clear how widely applicable the findings were. To address this, Slabbekoorn and

Ardie den Boer-Visser compared the songs sung by birds from ten pairs of populations, each pair comprising a population from a big city in Europe (including Amsterdam, London, Paris) and nearby rural sites. The songs of birds in urban populations were found to be shorter and sung faster than those of rural forest birds, and the minimum frequencies of the songs of urban birds were also higher. Changes in song in urban areas are probably essential for the birds to continue to maintain territories and mating opportunities above the din of city life. The animals are modifying the properties of their songs to avoid overlap and interference. Such a response is likely to be broadly a good thing, or at least an essential change on the part of the birds to allow them to continue communicating using sound. It shows that they can change, at least to some extent, to mitigate the presence of human noise. The caveat here is that there may be hidden costs with having to do this. Singing at a higher frequency may require more energy, and this may come at a long-term cost to the bird's survival. Having to expend more energy in singing may leave less for other things, such as mating or flight. Perhaps even more troubling is that research has shown that noise pollution can affect the species composition of bird communities, by affecting in some way how different species interact with one another. This means that assemblages of birds in urban areas may start to change, with some species becoming more or less common, and from this a multitude of potential knock-on effects for the ecosystem may be expected to follow.

The findings on bird song have been replicated in a wealth of other research, not just in birds but in a variety of other animals that communicate with sound. Animals change not just the properties of their song but the timing too. For instance, European blackbirds (*Turdus merula*) living near Madrid airport begin their dawn chorus earlier to avoid interference from planes.

Other birds, such as European robins, have been shown to sing more at night when noise during the day is high. In spite of potential hidden costs, it seems that many species can, and do, alter their song to avoid interference. Other terrestrial animals are not so fortunate, however. Not all bats hunt using echolocation. Some species eavesdrop on the rustling sounds made as their prey, which includes insects and other invertebrates, move through and over vegetation. These bats, which include the greater mouse-eared bat (Myotis myotis), have remarkably sensitive hearing. The problem is that the movements of prey have sound frequencies of about 10–15 kHz, and this is similar to the range of traffic noise; in the presence of roadside noise, bats take longer in searching for prey and they have reduced success at finding food.

The situation is not so different in the marine world, in part due to the considerable noise from shipping. Sometimes the sources of noise can be unexpected and unfortunate. For example, wildlife watching is often heralded as an important route to conserving species, and it clearly has many benefits in that regard. It provides income for local communities and an incentive for people to look after their natural resources. Yet ecotourism can in itself be costly for the animals. In some parts of the USA, at times, over twenty whale-watching boats have been reported following a single pod of orca (Orcinus orca), which communicate using echolocation to coordinate foraging and for a range of social interactions. Unfortunately, the orca are forced to make longer calls under the higher boat traffic noise.[12] The situation is similar for humpback whales, where males sing during mating. When males are exposed to playback sounds of low-frequency sonar, the males sing songs that are about one-third longer in duration to overcome this. Singing for longer likely has energy costs to the whales.

The effects of marine noise are widespread and arise even in unpredictable contexts. Animals such as shore crabs (*Carcinus maenas*), a species that is common all around the UK coastline, also suffer. This species is extremely common in many intertidal habitats, from tidal pools to mudflats, and relies on escaping predators with its camouflaged appearance and anti-predator behaviours, such as running away. The camouflage aspect is achieved in part by the crabs' ability to change how bright or dark they are over a period of weeks to better match the background. Individuals of this species, that we know of, do not primarily communicate using sound. It was a surprise then when research showed that they suffer disturbance to feeding, and are slower to retreat to shelter from a simulated predator when exposed to sounds of shipping.[13] This seems to be because the sounds are in some way stressful—crabs exposed to ship noise use up more oxygen than those that just encounter ambient noise. Not only that, but studies done in my own research lab, principally by Emily Carter, showed that when exposed to ship noise, crabs no longer change colour properly, and, as a result, they fail to merge with the background so effectively (Plate 10). It is a double whammy: crabs exposed to ship noise are both less camouflaged and less likely to run away when threatened. The same effects do not arise when crabs are exposed to natural noises, even when these are played as loudly as shipping sounds, so there must be something particularly stressful about the make-up of the machine noises.

Noise pollution can also affect the sensory abilities of an animal we encountered earlier that has highly specialized senses for a specific task: parasitoid flies. *Ormia* flies lay their eggs on cricket hosts, finding unwitting male crickets as they sing to attract females. The flies have a hearing system perfectly evolved to detect these male calls too. But recent work shows that the flies find it harder to locate singing crickets when in the presence of traffic noise.[14] In experiments in which speakers played the songs of male crickets, flies were less likely to locate speakers when they were in noisier surroundings. One might imagine that male crickets would relocate to places of higher noise to avoid falling prey to the flies, but the chances are that in doing so they might also be less able to attract a mate.

The rapid growth of human populations and the spread of urban areas and transport networks has led to another major source of pollution: light pollution. Long neglected, this problem is now attracting considerable attention. Some of this interest has stemmed from studies showing how the use of phones and tablets that emit lots of white or blue light can be bad for disrupting sleep rhythms in people. But much research by biologists has shown that light pollution can affect a host of biological processes in many organisms. In its simplest form, light pollution (in the scientific literature something often called 'artificial light at night') is the presence of light during nocturnal hours, when habitats would historically have been dark, with only light from the moon and stars (Figure 33). In fact, light pollution is more complex than the presence of night-time lighting alone because it also varies greatly in composition (wavelengths) and intensity. Specifically, in the past, much of the lighting used for cities and roadways was a kind of orange-yellow light, relatively dim in nature (most often in recent times in the form of low-pressure sodium streetlights). With the advent of LED lighting, with its greater energy efficiency, there has been a shift towards brighter, more intense light, and whiter light that covers more of the visible spectrum, including an abundance of shorter wavelengths (blue and green). All these components, the presence, intensity, and spectrum of lighting have been shown to influence animal physiology and behaviour.

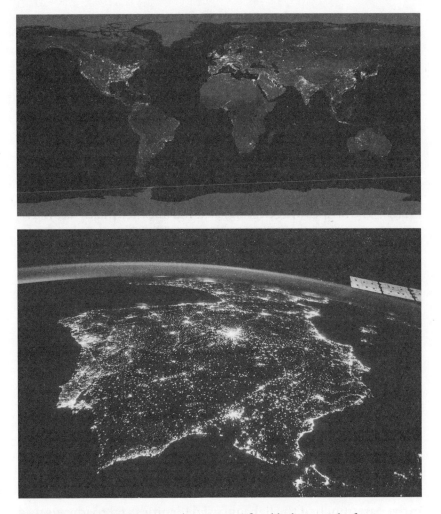

Figure 33. Remote imagery showing artificial light at night from satellite and observatory images across the world (top); and of Spain and Portugal at night from the International Space Station (below).

Among the earliest known negative impacts of light pollution on animals was the disorientation shown by hatchling turtles emerging on nesting beaches. As we have noted before, newly hatched turtles often head towards the sea, guided by the higher reflectance of light bouncing off the water surface compared to the beach. With the building of bars and restaurants and other developments at the back of many beaches, hatchings have long been observed to head in the wrong direction, moving to the back of the beach in particularly touristy locations.

Many of the recent investigations of light pollution, representing major steps forward in our understanding of its impact on animal behaviour and ecology, have been undertaken on moths. A good deal of this research has occurred at Wageningen University in the Netherlands. One study, by Frank van Langevelde and colleagues, showed that not only are many species of moth routinely attracted to artificial lights, but that the effect can be stronger for lights that emit spectra with shorter wavelengths (blue light).[15] Furthermore, these lights tended to attract moths that were larger in size and had larger eyes. As such, using lighting without these shorter wavelength emissions may be less detrimental.

Artificial lighting impacts moths in other, less expected ways too. For example, female winter moths (*Operophtera brumata*) are less likely to be found on trees when lights are present, and of those that do visit well-lit trees are much less likely to have mated than females that are found on dark trees. The results seem partly based on reduced female activity under lighting, but also changes in male behaviour. Males were less attracted to female pheromone lures under artificial lights. The reproductive behaviour of moths can therefore be affected by lighting, but so too it seems can moth responses to predators. For example, moths can become less than half as likely to take evasive action (such as diving to the ground) when they hear bat echolocation calls under LED-type street lights, compared to moths in darkness. As if all this were not enough, recent work has shown that moths show reduced feeding activity under light pollution, especially under shortwave lighting.

Long-term declines in moth populations are well appreciated and widely recorded, but they are not always well understood. A common suggestion is that artificial lighting at night may be one of the major drivers in such declines. Besides the implications for the moths themselves, a number of important knock-on effects arise from declining moth populations, not least in that moths and other nocturnal invertebrates form a major component of the diet of numerous predators, from bats to spiders. Declines in one group can often lead to declines in another. Potentially far-reaching effects are also being discovered elsewhere. Research published in 2017 showed that pollination of plants at night, including by moths, can be substantially compromised by light pollution.[16] One study showed that, in meadows with street lights nearby, this effect equated to plants being visited 60 per cent less often than those in darkness. This subsequently resulted in a 13 per cent reduction in fruit set, regardless of rates of day-time pollination. Another study found that moths carried pollen from twenty-eight or more species of plant, and so are important pollinators, but under light pollution (compared to dark sites) moths were less abundant near the ground, there were fewer species, and those present were found more often flying up near the lights. There was also some evidence for moths in light polluted areas carrying less pollen.

Arguably, the most far-reaching discoveries regarding light pollution and moths are those that show how light pollution seems to be playing a role in driving the contemporary evolution of urban populations. Florian Altermatt and Dieter Ebert in Switzerland took moth caterpillars from ten different populations, classed as either urban or rural, and reared them to adulthood.[17] Then they tested how the adult moths responded to artificial lighting, and discovered that moths that originated from urban populations were less likely to show attraction to lights. This is unlikely to be a behaviour they could have learnt or developed, because the moths originated from larvae that were reared in the laboratory under standardized conditions, regardless of the original population. Instead, the implication is that the moths in urban populations are showing genetic adaptations and evolving to be less responsive to lights. On the whole, this may be a good thing and shows how animals such as moths may adapt to cope with the modern world. But there may be costs, for example their being less active generally and less able to find food and mates.

Recent estimates of the proportion of the Earth's surface that is affected by light pollution range between about 10 per cent and 20 per cent of land area, and at least 0.2 per cent of the oceans. The latter may not sound much, but there is a great deal of intense light pollution around developed coastlines, beaches, ports, and industries, and coastal areas are often among the most productive for biodiversity. Fisheries can also represent major areas of artificial lighting. Light pollution has often been cited as spreading by more than 5 per cent globally per year, and so it is a rapidly accelerating problem. The issue is not just the extent of surface that light pollution covers, but the type of lighting used. As we have seen, older lights tended to be restricted in intensity and spectral range, but now there are major moves towards more intense and whiter lighting, especially via LEDs, which may be more energy efficient but can exacerbate the effects of light pollution. Another reason for the shift is public perception that people are kept safer with more night-time lighting. Yet there is little evidence that crime is lower with more street lighting, and if anything drivers may be less careful when driving under bright streetlights than in relative darkness. The solutions for managing light pollution therefore involve not only using less lighting but shifting public perspective regarding the need for excessive night-time lighting (and potentially a return to lights lacking certain wavelengths, though the effects here are complex). Using technology to better direct light to where it is needed and to limit its spread, or for smart lighting that only turns on when someone is nearby would no doubt help too.

Curiously, artificial night-time lighting is not detrimental to every species. For some, it may actually be of benefit. This is most obviously illustrated in orb-web spiders, which capture prey that lands in their webs. Many spiders lure prey to their webs with bright body colours, but flying insects need to be present in the environment in sufficient numbers in the first place for this to be successful. Given that many flying insects are attracted to lights, some spiders tend to be more successful in capturing prey if their web is located near lighting.[18] In fact, the spiders themselves show preferences for building webs close to light sources. This helps to explain why some spiders occur at higher density in urban areas, and have larger body sizes, than in rural locations. Not only that, but some spiders are also well known to occur more frequently along bridges, often making webs on the handrails, since these tend to be areas of high lighting. This does not necessarily mean that the spiders have evolved to respond to human environments: they may demonstrate such behaviour if in the past they were naturally attracted to build webs near water, where reflections of moonlight would increase insect numbers.

Differences in animal perceptions can also mean that we might not always appreciate sources of pollution and interference that are harmful. Take the potential impacts on some animals of being close to power lines. The cables carrying power across the landscape have been shown to 'spark' flashes of light in the ultraviolet, and this may be unpleasant for animals living in proximity to them, even potentially causing some birds or reindeer to avoid areas spanned by power lines. This may not seem like a bad thing, but if it affects movement or the migratory patterns of animals, or even simply reduces opportunities for foraging, then it may be quite harmful.

At the end of Chapter 7 we noted one of the curious findings in recent years involving the magnetic sense—that herding animals such as cattle and deer seem to orientate themselves with regards to the Earth's magnetic field. In the modern world, there are plenty of devices that could interfere with this, again not least power lines, since these are known to generate their own weak magnetic fields. Quite remarkably, the same team that showed evidence for orientation to the Earth's magnetic field in cattle and deer also later found evidence that the effect was abolished in herding animals in fields under or close to power lines (within 50–150 m); in such cases their orientation was random.[19] Furthermore, this effect declined with distance from the power lines, such that animals 500 m away from the lines aligned correctly on a north-south axis. This study further provides evidence that herding animals do respond to magnetic information, though as we have discussed before, the function of this response is unclear, so we cannot currently say how detrimental power lines might be.

More evidence of the potentially harmful impact on animals of electromagnetic noise resulting from our many technological devices and activities comes from studies of migrating birds. Much of this noise comes from AM radio signals and various electric devices found in homes and businesses (not power lines, as in the cattle example above, or mobile phones). Svenja Engels and Henrik Mouritsen at the University of Oldenburg and colleagues in 2014 tested how orientation behaviour by European robins was affected by electromagnetic noise.[20] After moving to a new university, the scientists were initially puzzled by why their birds could not orientate in their labs on campus in experiments that had worked well in the past. They investigated further. They determined how effectively the robins used their magnetic sense to orientate either in the presence of ambient electromagnetic noise or when kept in huts that were shielded from the noise by aluminium screens. Under the shielded conditions, the birds could orientate correctly, but when exposed to the noise their ability was abolished. These effects were found in urban but not rural locations, where the birds could consistently orientate correctly even without shielding. So far, there has not been much follow up work, though others have also suggested encountering similar issues when studying migrating animals. The implications are potentially substantial, in that the migratory behaviour of birds and other animals may be badly affected by such low levels of electromagnetic noise in urban worlds we have created. Birds have to migrate past and through urban areas all the time, so we can only imagine what this may be like for their senses.

In humans, studies have repeatedly failed to find detrimental effects from electromagnetic noise at industry-allowed levels in the environment. Those studies that have found impacts usually tend not to be replicated in follow-up work. But many species of animal have a far more refined ability to detect and respond to magnetic information, and so we should not be surprised that they may be affected even when we are not. The study of robins in particular was very robust, with multiple tests all being consistent and the experimenters blind to the treatments involved. While electromagnetic noise may have diverse effects, from potentially impeding the orientation behaviour of ants to contributing to population declines in migratory butterflies, this is another area where only time, and more research, can reveal the extent of any problems.

Any discussion of the impact that humans are having on the planet must at some point deal with climate change. The issue is widely reported, even if many humans have been extremely slow to react, such that there is little need for a discussion of what it is here, except to reiterate that climate change means a number of things. Most obviously, this includes changes in global and regional temperatures (air, ground, and water) and weather patterns, but it also alters the chemical composition of the oceans, snow and ice cover, and vegetation cover and composition in different geographical regions. In some regards, the impact of climate change on animal senses is hard to predict, and it may not always be direct, yet there are several areas where its influence may be pronounced.

One of the problems associated with increases in global emissions is that higher carbon dioxide levels in the atmosphere lead to more of the gas being absorbed by the oceans. When that happens it forms a weak acid, carbonic acid, and as a result, the oceans are becoming acidified, with a lower pH. Ultimately, the lowering of pH might affect the sense of smell in many animals, according to some research. This was the conclusion of a series of studies on clownfish, though more recent experiments have contested the findings.[21] The larvae of many marine fish spend time in the plankton as they initially develop; and, in a number of species, when they reach a certain developmental stage they travel towards their primary habitat, following chemical and sound cues. In much this way, larval orange clownfish (*Amphiprion percula*) use olfaction when they emerge from their planktonic larval stages and settle on suitable coral reefs. This is a charismatic and colourful orange and white species that associates with anemones, and is part of a group of fish made famous by the Disney film *Finding Nemo*. When larval clownfish were raised in the lab and tests made of their preferences to swim towards the smells of different stimuli, their responses changed under water conditions with acidity levels predicted for 2100. Here, the fish developed preferences for smells that could bring them into danger, and at still lower pH they failed to respond, as if their whole sense of smell no longer functioned properly. Similar findings have been replicated in other species, but another comprehensive study in 2020 failed to find any impact on fish responses with lowering pH, so this is a matter of current debate. Nonetheless, other work has also shown that changes in pH can alter the way that the olfactory systems and receptors of fish actually work. If the clownfish work remains robust, then findings such as those are clearly troubling. They suggest that clownfish and many other species that rely on olfaction for finding suitable habitats and avoiding major risks may be badly affected by changing ocean properties. Perhaps it is a fool's hope, but hopefully we won't have to find out the real answer in the future outside of the laboratory.

We should bear in mind too that these pressures must be placed against a backdrop of a host of other issues, including noise pollution, that may also make finding reef habitats a challenge. On that note, interestingly, recent work has shown that juvenile fish responses to the sounds of coral reefs can be used to help restore damaged habitats. Coral reefs tend to be very noisy places. The inhabitants make sounds for communication and simply as they go about their daily lives, and many larval fish use sound as well as chemical cues to find suitable places to settle. Unfortunately, in addition to rising sea temperatures causing coral bleaching, physical damage to many reefs is leading to large swathes of environment being degraded and abandoned. Recent studies have shown that when the sounds of healthy reefs are played back in the vicinity of degraded coral rubble patches, a variety of juvenile fish are attracted.[22] In fact, twice as many fish, including grazers and predators, settle compared to patches that are not accompanied by coral sounds. This 'acoustic enrichment' can't stop the effects of human actions destroying reefs in the first place, but it may be a valuable tool in helping to restore reefs.

Beyond olfaction, climate change can have other more subtle effects related to animal senses and how they are used. For example, as temperatures and weather patterns change, there will in many locations be a reduction in the extent and duration of winter snow cover. From ptarmigan to artic fox, a number of animals change colour to a white appearance in winter for camouflage against the snow, to be hidden from the eyes of predators and prey. In snowshoe hares in North America, declines in snowpack extent and duration are already resulting in an increased number of days when hares do not match the background properly.[23] The hares change to a white appearance but either the snow has not yet arrived or it has already melted. This has serious implications for survival, with more hares being eaten by predators. We are only just beginning to understand their nature, but changes to the visual appearance of habitats via climate change, coupled with problems such as light pollution, will have major impacts on many animals' vision, their appearance, and the interaction between the two.

There is a risk that in all this discussion of how humans are affecting animal senses we might do little more than spread doom and despair. Certainly, there is no getting around the fact that we are thoroughly damaging the planet, and most of the work in this area and on animal senses focuses on what is going wrong. Solutions are rather harder to arrive at. This is partly because the solutions cannot come from science alone, but from politicians, industries, and the public. Let's end the book adopting a more positive angle, and consider how knowledge of animal senses has been put to good use in seeking to solve one of the problems of our own creation.

Issues caused by overfishing and capture methods in our oceans are well documented (though not always acted upon). Many target species are heavily overfished at unsustainable levels, and methods such as bottom trawling cause large environmental impacts through habitat destruction. Another problem with fishing is bycatch: many species are caught that are not the target species, with staggering numbers being thrown back or discarded, and ultimately dying. Not only does this include species similar to those targeted, but a host of other animals, from sharks and dolphins to turtles that are often protected and endangered in their own right. Once non-target species are caught, their chances of survival even if released are often slim, so the best way to solve the problem is to prevent bycatch from happening in the first place.

Sea turtles are often caught in fishing nets, becoming tangled with the gear and unable to escape. The numbers are alarming, with 250,000 loggerhead and leatherback turtles estimated to have been captured by longline fisheries in 2000 alone. So how to stop this? One approach that has been investigated since the early 2000s has been to use either visual deterrents in nets, or devices to make the nets more visible to turtles.[24] Visual information is valuable to turtles in prey identification and capture, and they are known to have very good colour vision across a wide spectrum of wavelengths, not dissimilar to that of birds. By contrast, many open ocean marine fish, especially those in deeper waters, have relatively poor or restricted colour vision, and so using coloured stimuli is a fruitful avenue of research.

Bait that had been dyed red or blue seemed to have no clear effect on turtle preferences, whereas using lights seems to be a more promising avenue. Research led by John Wang at the

University of Hawaii (alongside Kenneth Lohmann) reported results testing the responses of loggerhead turtles to lightsticks, which are sometimes used in longline fisheries (with hooked lines running over tens of kilometres) to attract tuna and sword-fish, and the lines can capture turtles too. The work showed that turtles were attracted to green, blue, yellow, and orange lights, potentially meaning that in the wild they may be attracted to them along fishing lines. But not all results are consistent. Research shortly afterwards on leatherback turtles found little evidence of orientation towards lights, and even some evidence that they orientated away from them. This hits upon an early problem: if the responses of different species to coloured lights may be either one of attraction or repulsion, causing the wrong response may in fact make matters worse for turtle bycatch.

These early studies had been conducted in a lab environment, where light conditions would be very different to that of the field. Later work addressed this. When gill nets in Mexico were modi-fied to contain shark shapes (a predator of turtles), battery powered green LED lights, or green chemical lightsticks, there were striking differences in how effectively these prevented the capture of green sea turtles and target fish. Adding shark shapes to nets was very effective in preventing turtle capture, reducing capture by over 50 per cent. However, the downside was that the silhouettes also decreased the capture of the target fish species by 45 per cent. Clearly this then is not viable for commercial fisher-ies. In comparison, nets containing LED lights or lightsticks reduced turtle capture by 40 per cent and 60 per cent respect-ively, with little impact on fish.

Wang and others also tested the responses of green sea turtles to ultraviolet emitting LED lights. This is potentially especially promising because turtles can see UV light, yet many target spe-cies cannot, meaning that the turtles may see and avoid the nets, unlike many target fish. Experiments with gill nets showed a 40 per cent reduction in turtle capture in nets with the UV LEDs compared to nets without them, while the addition of the LEDs did not affect target fish capture. Whether or not this was because the turtles actively avoid UV lights, or because the UV lights alert them to the net, is as yet unknown. Further recent work on green turtles and small-scale fisheries has also found green LED lights to be an effective way of reducing turtle capture from gill nets (Plate 11).

The use of LED lights in fisheries as a way of preventing bycatch, especially of highly visual animals such as turtles, is very promising, and target species appear to be relatively unaffected. But, as always, things are never simple and there are other considerations too. For example, using short-term lightsticks may produce lots of unwanted plastic waste, and LED lights may be too expensive for low-cost local fisheries. In addition, there may be cause for concern regarding whether such devices may add yet another damaging source of light pollution. However, the costs of LED lights are generally low and declining, and using solar powered ones may further reduce costs while eliminating battery waste. What this example shows is that exploiting the sensory biology of different species can help to solve some of the major problems our food demands are creating.

So diverse and often so wonderfully refined are animal senses that conveying the vast and impressive range of sensory systems in nature is a challenge. It is also hard to express just how precise and accurate these senses can be across species, given the intricate levels of detail to which many animals respond. There is a risk that doing so would end up as a list of facts and figures regarding what different animals can perceive and detect, losing sight of the bigger picture. At the start of this book, I outlined why animal senses in nature should be so diverse. One of the main reasons is that animals have been moulded by millions of years of evolution to live and perform a variety of tasks effectively in the environments in which they reside. The information available for the senses to encode, and the habitats in which each animal lives, shapes the senses an animal has and how they work. Evolution is also efficient and careful with how resources are allocated, and sensory systems are very costly. Rather than seeing countless species all around us, each with every single one of their senses being the pinnacle of what is possible, we instead observe that evolution and development has honed those senses that the animal needs most, and scaled back on the others. Choosing one animal that has the greatest or most impressive senses is futile, because what really matters is how each animal uses its own senses to find food, court with mates, navigate, and avoid predators. Different species have arrived at a plethora of wonderful, diverse, and sophisticated solutions to match these challenges. And very different animals faced with the same challenges have sometimes independently converged on the same or similar solutions. Ultimately, animals simply cannot interact with the environment without their senses, and these are critical for anything from day-to-day life to enabling the yearly migrations of birds over thousands of kilometres, or precise behaviours such as locating, identifying, and capturing a small flying moth on the wing in pitch darkness.

Uncovering the secrets of animal senses has required studies ranging from molecular biology and neuroscience, through to behaviour and ecology. We have learnt a lot about how they work, yet in other ways we are just scratching the surface. The senses of other organisms, such as plants and fungi, are even more mysterious and hard to fathom. Yet they are well known to perceive their world in a variety of ways too, from detecting light to chemicals. As one example, recent research has shown that the flowers of some plants can even respond to specific frequencies of mechanical stimulation caused by the sounds of flying pollinator insects. Within minutes of sensing these sounds, plants increase the sugar concentration of their nectar.[25]

The challenge of knowing how animals, and other organisms more widely, perceive the world is no doubt helped by modern technology and clever experiments, but there is much left to learn and discover. The even bigger task of facing up to the challenges of the modern world, mostly of our own creation, is urgent. We have to hope, and act quickly, so that we still have the opportunity to understand nature's secret worlds, both for the fundamental advancement of knowledge itself and for the wider benefits which that understanding can bring. The senses we ourselves possess are far from remarkable, but we have something unparalleled in evolution: our deep level of conscious awareness. It is our responsibility to make sure we use that for the good of all life on Earth.

NOTES AND REFERENCES

CHAPTER 1

1. LOHMANN, K. J., PENTCHEFF, N. D., NEVITT, G. A., STETTEN, G. D., ZIMMER-FAUST, R. K., JARRARD, H. E., and BOLES, L. C. 1995. Magnetic orientation of spiny lobsters in the ocean: experiments with undersea coil systems. *Journal of Experimental Biology* 198:2041–8.
   BOLES, L. C., and LOHMANN, K. J. 2003. True navigation and magnetic maps in spiny lobsters. *Nature* 421:60–3.
2. GRACHEVA, E. O., CORDERO-MORALES, J. F., GONZÁLEZ-CARCACÍA, J. A., INGOLIA, N. T., MANNO, C., ARANGUREN, C. I., WEISSMAN, J. S., and JULIUS D. 2011. Ganglion-specific splicing of TRPV1 underlies infrared sensation in vampire bats. *Nature* 476:88–92.
3. ROBERT, D., AMOROSO, J., and HOY, R. R. 1992. The evolutionary convergence of hearing in a parasitoid fly and its cricket host. *Science* 258:1135–7.
4. CROSS, F. R., and JACKSON, R. R. 2011. Olfaction-based anthropophily in a mosquito-specialist predator. *Biology Letters* 7:510–12.
5. NARENDRA, A., REID, S. F., GREINER, B., PETERS, R. A., HEMMI, J. M., RIBI, W. A., and ZEIL, J. 2011. Caste-specific visual adaptations to distinct daily activity schedules in Australian *Myrmecia* ants. *Proceedings of the Royal Society B: Biological Sciences* 278:1141–9.
6. ZHAO, H., LI, J., and ZHANG, J. 2015. Molecular evidence for the loss of three basic tastes in penguins. *Current Biology* 25:R141–R142.
7. LAUGHLIN, S. B., DE RUYTER VAN STEVENINCK, R. R., and ANDERSON, J. C. 1998. The metabolic cost of neural information. *Nature Neuroscience* 1:36–41.
8. TAN, S., AMOS, W., and LAUGHLIN S. B. 2005. Captivity selects for smaller eyes. *Current Biology* 15:R540–R542.
9. MORAN, D., SOFTLEY, R., and WARRANT, E. J. 2015. The energetic cost of vision and the evolution of eyeless Mexican cavefish. *Science Advances* 1:e1500363.

10. BLEST, A. D. 1978. The rapid synthesis and destruction of photo-receptor membrane by a dinopid spider: a daily cycle. *Proceedings of the Royal Society of London. Series B, Biological Sciences* 200:463–83.

11. VAN BREUGEL, F., RIFFELL, J., FAIRHALL, A., and DICKINSON, M. H. 2015. Mosquitoes use vision to associate odor plumes with thermal targets. *Current Biology* 25:2123–9.

    CARDÉ, R. T. 2015. Multi-cue integration: how female mosquitoes locate a human host. *Current Biology* 25:R793–R795.

12. GRACHEVA, E. O., INGOLIA, N. T., KELLY, Y. M., CORDERO-MORALES, J. F., HOLLOPETER, G., CHESLER, A. T., SÁNCHEZ, E. E., PEREZ, J. C., WEISSMAN, J. S., and JULIUS, D. 2010. Molecular basis of infrared detection by snakes. *Nature* 464:1006–12.

    GENG, J., LIANG, D., JIANG, K., and ZHANG, P. 2011. Molecular evolution of the infrared sensory gene TRPA1 in snakes and implications for functional studies. *PLoS ONE* 6:e28644.

    YOKOYAMA, S., ALTUN, A., and DeNARDO, D. F. 2011. Molecular convergence of infrared vision in snakes. *Molecular Biology and Evolution* 28:45–8.

13. YOSHIDA, M., ITOH, Y., ÔMURA, H., ARIKAWA, K., and KINOSHITA, M. 2015. Plant scents modify innate colour preference in foraging swallowtail butterflies. *Biology Letters* 11:20150390.

CHAPTER 2

1. MOIR, H. M., JACKSON, J. C., and WINDMILL, J. F. C. 2013. Extremely high frequency sensitivity in a 'simple' ear. *Biology Letters* 9:20130241.

2. PAYNE, R. S., and DRURY, W. H. 1958. *Tyto alba*, Part II. *Natural History NY* 67:316–23.

    PAYNE, R. S. 1962. How the barn owl locates prey by hearing. In *The Living Bird*. 1st Annual, Ithacat, NY: Cornell Lab. of Ornithology, 151–9.

    PAYNE, R. S. 1971. Acoustic location of prey by barn owls (*Tyto alba*). *Journal of Experimental Biology* 54:535–73.

3. KNUDSEN, E. I., and KONISHI, M. 1978. Center-surround organization of auditory receptive fields in the owl. *Science* 202:778–80.

KNUDSEN, E. I., and KONISHI, M. 1979. Mechanisms of sound localization in the barn owl (Tyto alba). Journal of Comparative Physiology A 133:13–21.

KNUDSEN, E. I., and KONISHI, M. 1979. Mechanisms of sound localization in the barn owl (Tyto alba). Journal of Comparative Physiology A 133:13–21.

KNUDSEN, E. I., BLASDEL, G. G., and KONISHI, M. 1979. Sound localisation by the barn owl (Tyto alba) measured with the search coil technique. Journal of Comparative Physiology A 133:1–11.

KONISHI, M. 1973. How the owl tracks its prey: experiments with trained barn owls reveal how their acute sense of hearing enables them to catch prey in the dark. American Scientist 61:414–24.

4. CARR, C. E., and KONISHI, M. 1990. A circuit for detection of interaural time differences in the brain stem of the barn owl. Journal of Neuroscience 10:3227–46.

ASHIDA, G., ABE, K., FUNABIKI, K., and KONISHI, M. 2007. Passive soma facilitates submillisecond coincidence detection in the owl's auditory system. Journal of Neuroscience 97:2267–82.

5. KRUMM, B., KLUMP, G., KÖPPL, C., and LANGEMANN, U. 2017. Barn owls have ageless ears. Proceedings of the Royal Society of London. Series B 284:20171584.

6. ANDERSON, J. W. 1954. The production of ultrasonic sounds by laboratory rats and other mammals. Science 119:808–9.

NOIROT, E. 1966. Ultra-sounds in young rodents. I. Changes with age in albino mice. Animal Behaviour 14:459–62.

NOIROT, E. 1968. Ultrasounds in young rodents II. Changes with age in albino rats. Animal Behaviour 16:129–34.

SEWELL, G. D. 1967. Ultrasound in adult rodents. Nature 215:512.

SEWELL, G. D. 1968. Ultrasound in rodents. Nature 217:682–3.

SEWELL, G. D. 1970. Ultrasonic communication in rodents. Nature 227:410.

SEWELL, G. D. 1970. Ultrasonic signals from rodents. Ultrasonics 8:26–30.

7. HOLY, T. E., and GUO, Z. 2005. Ultrasonic songs of male mice. PLoS ONE 3:e386.

8. As always, the detailed picture is a little more complex. For example, it was shown soon after (Burgdorf et al. 2008) that 50 kHz calls actually comprise two types: a frequency modulated type that varies somewhat in frequency as it is produced and a flatter constant frequency version. The former seems to relate to directly positive experiences in rats, such as play and mating, whereas the latter is more related to neutral circumstances and aggression. Work in 2010 (Takahashi et al. 2010) also found evidence that the 50 kHz call may comprise two calls: a 40 and a 60 kHz call.

BURGDORF, J., KROES, R. A., and MOSKAL, J. R. 2008. Ultrasonic vocalizations of rats (*Rattus norvegicus*) during mating, play, and aggression: behavioral concomitants, relationship to reward, and self-administration of playback. *Journal of Comparative Physiology A* 122:357–67.

TAKAHASHI, N., KASHINO, M., and HIRONAKA, N. 2010. Structure of rat ultrasonic vocalizations and Its relevance to behavior. *PLoS ONE* 5:e14115.

9. RYGULA, R., PLUTA, H., and POPIK, P. 2012. Laughing rats are optimistic. *PLoS ONE* 7:e51959.

10. OKOBI, D. E., BANERJEE, A., MATHESON, A. M. M., PHELPS, S. M., and LONG, M. A. 2019. Motor cortical control of vocal interaction in neotropical singing mice. *Science* 363:983–8.

11. LAZURE, L., and FENTON, M. B. 2010. High duty cycle echolocation and prey detection by bats. *Journal of Experimental Biology* 214:1131–7.

SCHNITZLER, H. U., and DENZINGER, A. 2011. Auditory fovea and Doppler shift compensation: adaptations for flutter detection in echolocating bats using CF-FM signals. *Journal of Comparative Physiology A* 197:541–59.

12. SULLIVAN, W. E., 3rd. 1982. Neural representation of target distance in auditory cortex of the echolocating bat *Myotis lucifugus. Journal of Neurophysiology* 48:1011–32.

SULLIVAN, W. E., 3rd. 1982. Possible neural mechanisms of target distance coding in auditory system of the echolocating bat *Myotis lucifugus. Journal of Neurophysiology* 48:1033–47.

13. SUGA, N., SIMMONS, J. A., and JEN, P. H.-S. 1975. Peripheral specialization for fine analysis of doppler-shifted echoes in the auditory system

of the 'CF-FM' bat *Pteronotus parnellii*. *Journal of Experimental Biology* 63:161–92.

SUGA, N., and JEN, P. H.-S. 1977. Further studies on the peripheral auditory system of 'CF-FM' bats specialized for fine frequency analysis of doppler-shifted echoes. *Journal of Experimental Biology* 69:207–32.

POLLAK, G. D., and BODENHAMER, R. D. 1981. Specialized characteristics of single units in inferior colliculus of mustache bat: frequency representation, tuning, and discharge patterns *Journal of Neurophysiology* 46:605–620.

14. LI, G., WANG, J., ROSSITER, S. J., JONES, G., COTTON, J. A., and ZHANG, S. 2008. The hearing gene *Prestin* reunites echolocating bats. *PNAS* 105:13959–64.

15. LIU, Y., COTTON, J. A., SHEN, B., HAN, X., ROSSITER, S. J., and ZHANG, S. 2010. Convergent sequence evolution between echolocating bats and dolphins. *Current Biology* 20:R53–R54.

LI, Y., LIU, Z., SHI, P., and ZHANG, J. 2010. The hearing gene *Prestin* unites echolocating bats and whales. *Current Biology* 20:R55–R56.

CHAPTER 3

1. DACKE, M., BAIRD, E., BYRNE, M., SCHOLTZ, C. H., and WARRANT, E. J. 2013. Dung beetles use the Milky Way for orientation. *Current Biology* 23:298–300.

FOSTER, J. J., EL JUNDI, B., SMOLKA, J., KHALDY, L., NILSSON, D.-E., BYRNE, M. J., and DACKE, M. 2017. Stellar performance: mechanisms underlying Milky Way orientation in dung beetles. *Philosophical Transactions of the Royal Society B* 372:20160079.

2. NEWTON, I. 1718. *Opticks, or, A treatise of the reflections, refractions, inflections and colours of light*, 2nd edn, with additions. London: Printed for W. and J. Innys.

3. WALLACE, A. R. 1889. *Darwinism: An Exposition of the Theory of Natural Selection with Some of its Applications*. London: Macmillan & Co.

4. TURNER, C. H. 1910. Experiments on color-vision of the honey bee. *Biological Bulletin* 19:257–79.

SEE ALSO:

GIURFA, M., and DE BRITO SANCHEZ, M. G. 2020. Black lives matter: revisiting Charles Henry Turner's experiments on honey bee color vision. *Current Biology* 30:R1235–9.

5. LI, D., and LIM, M. L. M. 2005. Ultraviolet cues affect the foraging behaviour of jumping spiders. *Animal Behaviour* 70:771–6.

6. LIM, M. L. M., and LI, D. 2006. Extreme ultraviolet sexual dimorphism in jumping spiders (Araneae: Salticidae). *Biological Journal of the Linnean Society* 89:397–406.

7. LI, J., LIM, M. L. M., ZHANG, Z., LIU, Q., LIU, F., CHEN, J., and LI, D. 2008. Sexual dichromatism and male colour morph in ultraviolet-B reflectance in two populations of the jumping spider *Phintella vittata* (Araneae: Salticidae) from tropical China. *Biological Journal of the Linnean Society* 94:7–20.

   LI, J., ZHANG, Z., LIU, F., LIU, Q., GAN, W., CHEN, J., and LIM, M. L. M. 2008. UVB-based mate-choice cues used by females of the jumping spider *Phintella vittata*. *Current Biology* 18:699–703.

8. LIM, M. L. M., LAND, M. F., and LI, D. 2007. Sex-specific UV and fluorescence signals in jumping spiders. *Science* 315:481.

9. LUBBOCK, SIR J. 1881. Observations on ants, bees, and wasps. VIII. Experiments with light of different wavelengths. *Journal of the Linnean Society of London* 15:362–87.

10. MADDOCKS, S. A., CHURCH, S. C., and CUTHILL, I. C. 2001. The effects of the light environment on prey choice by zebra finches. *Journal of Experimental Biology* 204:2509–15.

11. To our eyes, the Ambon damselfish (*Pomacentrus amboinensis*) and Lemon damselfish (*P. moluccensis*) look very similar—a rich golden yellow colour. There's not much that looks different between them. However, male ambon damselfish can somehow tell rival males of their own species from males of the lemon damselfish. They need to tell them apart because male ambons defend territories on the reef, and generally the main threat (such as for competition for females) is from males of the same species. That means that an ambon male should be more aggressive to the ambon rival males than males from the lemon damselfish species. In this instance, if you take

photographs of the damselfish under UV light, you find that the ambon damsel fish have highly elaborate face patterns that differ between individuals. Experiments placing either ambon or lemon damselfish males in the territory of ambon males have shown that the territory owners only show aggression directed to males of the same species when UV light, and hence the face patterns could be seen. When the face patterns were lost, the territory owners did not distinguish between males of either species.

SIEBECK, U. E., PARKER, A. N., SPRENGER, D., MATHGER, L. M., and WALLIS, G. 2010. A species of reef fish that uses ultraviolet patterns for covert face recognition. *Current Biology* 20:407–10.

12. PEARN, S. M., BENNETT, A. T. D., and CUTHILL, I. C. 2001. Ultraviolet vision, fluorescence and mate choice in a parrot, the budgerigar *Melopsittacus undulatus*. *Proceedings of the Royal Society B* 268:2273–9.

ARNOLD, K. E., OWENS, I. P. F., and MARSHALL, N. J. 2002. Fluorescent signaling in parrots. *Science* 295:92.

13. GRUBER, D. F., LOEW, E. R., DEHEYN, D. D., AKKAYNAK, D., GAFFNEY, J. P., SMITH, W. L., DAVIS, M. P., STERN, J. H., PIERIBONE, V. A., and SPARKS, J. S. 2016. Biofluorescence in catsharks (Scyliorhinidae): fundamental description and relevance for elasmobranch visual ecology. *Scientific Reports* 6:24751.

BITTON, P.-P., HARANT, U. K., FRITSCH, R., CHAMP, C. M., TEMPLE, S. E., and MICHIELS, N. K. 2017. Red fluorescence of the triple fin *Tripterygion delaisi* is increasingly visible against background light with increasing depth. *Royal Society Open Science* 4:161009.

GERLACH, T., SPRENGER, D., and MICHIELS, N. K. 2014. Fairy wrasses perceive and respond to their deep red fluorescent coloration. *Proceedings of the Royal Society of London*. Series B 281:20140787.

HENDRY, T. A., LIGON, R. A., BESLER, K. R., FAY, R. L., and SMEE, M. R. 2018. Visual detection and avoidance of pathogenic bacteria by aphids. *Current Biology* 28:3158–64.

TABOADA, C., BRUNETTI, A. E., PEDRON, F. N., NETO, F. C., ESTRIN, D. A., BARI, S. E., CHEMES, L. B., LOPES, N. P., LAGORIO, M. G., and FAIVOVICH, J. 2017. Naturally occurring fluorescence in frogs. *PNAS* 114:3672–7.

14. MAZEL, C. H., CRONIN, T. W., CALDWELL, R. L., and MARSHALL, N. J. 2004. Fluorescent enhancement of signalling in a mantis shrimp. *Science* 303:51.

15. MARSHALL, N. J. 1988. A unique colour and polarization vision system in mantis shrimps. *Nature* 333:557–60.

    CRONIN, T. W., and MARSHALL, N. J. 1989. Multiple spectral classes of photoreceptors in the retinas of gonodactyloid stomatopod crustaceans. *Journal of Comparative Physiology A* 166:261–75.

    CRONIN, T. W., and MARSHALL, N. J. 1989. A retina with at least ten spectral types of photoreceptors in a mantis shrimp. *Nature* 339:137–40.

    MARSHALL, J., and OBERWINKLER, J. 1999. The colourful world of the mantis shrimp. *Nature* 401:873–4.

16. THOEN, H. H., HOW, M. J., CHIOU, T.-H., and MARSHALL, J. 2014. A different form of color vision in mantis shrimp. *Science* 343:411–13.

17. MARSHALL, J., CRONIN, T. W., SHASHAR, N., and LAND, M. 1999. Behavioural evidence for polarisation vision in stomatopods reveals a potential channel for communication *Current Biology* 9:755–8.

18. CHIOU, T.-H., KLEINLOGEL, S., CRONIN, T., CALDWELL, R., LOEFFLER, B., SIDDIQI, A., GOLDIZEN, A., and MARSHALL, J. 2008. Circular polarization vision in a stomatopod crustacean. *Current Biology* 18:429–34.

    GAGNON, Y. L., MARIE, R., HOW, M. J., JUSTIN, N., GAGNON, Y. L., TEMPLIN, R. M., HOW, M. J., and MARSHALL, N. J. 2015. Circularly polarized light as a communication signal in mantis shrimps. *Current Biology* 25:3074–8.

19. LAND, M. 2008. Biological optics: circularly polarised crustaceans. *Current Biology* 18:R348–R349.

20. CRONIN, T. W., CALDWELL, R. L., and MARSHALL, J. 2001. Tunable colour vision in a mantis shrimp. *Nature* 411:547–8.

21. EWERT, J.-P., and HOCK, F. 1972. Movement-sensitive neurones in the toad's retina. *Experimental Brain Research* 16:41–59.

    BECK, A., and EWERT, J.-P. 1979. Prey selection by toads (*Bufo bufo* L.) in response to configurational stimuli moved in the visual field z,y-coordinates. *Journal of Comparative Physiology A* 129:207–9.

22. BORCHERS, H.-W., and EWERT, J.-P. 1979. Correlation between behavioral and neuronal activities of toads *Bufo bufo* (L.) in response to moving configurational prey stimuli. *Behavioural Processes* 4:99–106.

EWERT, J.-P., BORCHERS, H.-W., and WIETERSHEIM, A. V. 1978. Question of prey feature detectors in the toad's *Bufo bufo* (L.) visual system: a correlation analysis. *Journal of Comparative Physiology A* 126:43–7.

CHAPTER 4

1. KALMIJN, A. J. 1971. The electric sense of sharks and rays. *Journal of Experimental Biology* 55:371–83.

2. KEMPSTER, R. M., HART, N. S., and COLLIN, P. 2013. Survival of the stillest: predator avoidance in shark Embryos. *PLoS ONE* 8:e52551.

BEDORE, C. N., KAJIURA, S. M., and JOHNSEN, S. 2016. Freezing behaviour facilitates bioelectric crypsis in cuttlefish faced with predation risk. *Proceedings of the Royal Society of London. Series B* 282:20151886.

3. WU, C. H. 1984. Electric fish and the discovery of animal electricity. *American Scientist* 72:598–607.

4. LISSMANN, H. W. 1951. Continuous electrical signals from the tail of a fish, *Gymnarchus niloticus* Cuv. *Nature* 167:201–2.

LISSMANN, H. W. 1958. On the function and evolution of electric organs in fish. *Journal of Experimental Biology* 35:156–91.

LISSMANN, H. W., and MACHIN, K. E. 1958. The mechanism of object location in *Gymnarchus niloticus* and similar fish. *Journal of Experimental Biology* 35:451–86.

5. KNUDSEN, E. I. 1975. Spatial aspects of the electric fields generated by weakly electric fish. *Journal of Comparative Physiology A* 99:103–18.

6. LAVOUÉ, S., MIYA, M., ARNEGARD, M. E., SULLIVAN, J. P., HOPKINS, C. D., and NISHIDA, M. 2012. Comparable Ages for the independent origins of electrogenesis in African and South American weakly electric fishes. *PLoS ONE* 7:e36287.

7. ZAKON, H. H., LU, Y., ZWICKL, D. J., and HILLIS, M. 2006. Sodium channel genes and the evolution of diversity in communication signals of electric fishes: convergent molecular evolution. *PNAS* 103:3675–80.

ARNEGARD, M. E., ZWICKL, D. J., and ZAKON, H. H. 2010. Old gene duplication facilitates origin and diversification of an innovative communication system—twice. *PNAS* 107:22172–7.

8. HOPKINS, C. D. 1972. Sex differences in electric signaling in an electric fish. *Science* 176:1035–7.

9. HAGEDORN, M., and HEILIGENBERG, W. 1985. Court and spark: electric signals in the courtship and mating of gymnotoid fish. *Animal Behaviour* 33:254–65.

10. CURTIS, C. C., and STODDARD, P. K. 2003. Mate preference in female electric fish, *Brachyhypopomus pinnicaudatus*. *Animal Behaviour* 66:329–36.

11. SALAZAR, V. L., and STODDARD, P. K. 2008. Sex differences in energetic costs explain sexual dimorphism in the circadian rhythm modulation of the electrocommunication signal of the gymnotiform fish *Brachyhypopomus pinnicaudatus*. *Journal of Experimental Biology* 211:1012–20.

12. SCHEICH, H., LANGNER, G., TIDEMANN, C., COLES, R. B., and GUPPY, A. 1986. Electroreception and electrolocation in platypus. *Nature* 319:401–2.

ASAHARA, M., KOIZUMI, M., MACRINI, T. E., HAND, S. J., and ARCHER, M. 2016. Comparative cranial morphology in living and extinct platypuses: feeding behavior, electroreception, and loss of teeth. *Science Advances* 2:e1601329.

GREGORY, J. E., IGGO, A., McINTYRE, A. K., and PROSKE, U. 1987. Electroreceptors in the platypus. *Nature* 326:386–7.

GREGORY, J. E., IGGO, A., McINTYRE, A. K., and PROSKE, U. 1988. Receptors in the bill of the platypus. *Journal of Physiology* 400:349–66.

GREGORY, J. E., IGGO, A., McINTYRE, A. K., and PROSKE, U. 1989. Responses of electroreceptors in the platypus bill to steady and alternating potentials. *Journal of Physiology* 408:391–404.

13. GREGORY, J. E., IGGO, A., McINTYRE, A. K., and PROSKE, U. 1989. Responses of electroreceptors in the snout of the echidna. *Journal of Physiology* 414:521–38.

14. CLARKE, D., WHITNEY, H., SUTTON, G., and ROBERT, D. 2013. Detection and learning of floral electric fields by bumblebees. *Science* 340:66–9.

CHAPTER 5

1. HANKE, W., WIESKOTTEN, S., MARSHALL, C., and DEHNHARDT, G. 2013. Hydrodynamic perception in true seals (Phocidae) and eared seals (Otariidae). *Journal of Comparative Physiology A* 199:421–40.

   NIESTEROK, B., KRÜGER, Y., WIESKOTTEN, S., DEHNHARDT, G., and HANKE, W. 2017. Hydrodynamic detection and localization of artificial flatfish breathing currents by harbour seals (*Phoca vitulina*). *Journal of Experimental Biology* 220:174–85.

2. HAMILTON, W. J. 1931. Habits of the star-nosed mole, *Condylura cristata. Journal of Mammalogy* 12:345–55.

3. CATANIA, K. C., and KAAS, J. H. 1996. The unusual nose and brain of the star-nosed mole. *BioScience* 46:578–86.

   CATANIA, K. C., and REMPLE, F. E. 2004. Tactile foveation in the star-nosed mole. *Brain, Behaviour and Evolution* 63:1–12.

   CATANIA, K. C. 1995. Structure and innervation of the sensory organs on the snout of the star-nosed mole. *Journal of Comparative Neurology* 351:536–48.

4. CATANIA, K. C., NORTHCUTT, G. R., KAAS, J. H., and BECK, P. D. 1993. Nose stars and brain stripes. *Nature* 364:493.

   CATANIA, K. C. 1996. Ultrastructure of the Eimer's organ of the star-nosed mole. *Journal of Comparative Neurology* 365:343–54.

   CATANIA, K. C. 2011. The sense of touch in the star-nosed mole: from mechanoreceptors to the brain. *Philosophical Transactions of the Royal Society B* 366:3016–25.

   CATANIA, K. C. 2005. Star-nosed moles. *Current Biology* 15:R863–R864.

5. CRISH, S. D., COMER, C. M., MARASCO, P. D., and CATANIA, C. 2003. Somatosensation in the superior colliculus of the star-nosed mole. *Journal of Comparative Neurology* 464:415–25.

6. CATANIA, K. C., and KAAS, J. H. 1995. Organization of the somatosensory cortex of the star-nosed mole. *Journal of Comparative Neurology* 351:549–67.

   CATANIA, K. C., and KAAS, J. H. 1997. Organization of somatosensory cortex and distribution of corticospinal neurons in the eastern mole (*Scalopus aquaticus*). *Journal of Comparative Neurobiology* 378:337–353.

CATANIA, K. C., and KAAS, J. H. 1997. Somatosensory fovea in the star-nosed mole: behavioral use of the star in relation to innervation patterns and cortical representation. *Journal of Comparative Neurology* 387:215–33.

7. CATANIA, K. C. 2000. Cortical organization in moles: evidence of new areas and a specialized S2. *Somatosensory & Motor Research* 17:335–47.

CATANIA, K. C. 2011. The sense of touch in the star-nosed mole: from mechanoreceptors to the brain. *Philosophical Transactions of the Royal Society B* 366:3016–25.

8. CATANIA, K. C., and REMPLE, M. S. 2002. Somatosensory cortex dominated by the representation of teeth in the naked mole-rat brain. *PNAS* 99:5692–7.

9. JACKSON, K., BUTLER, D. G., and YOUSON, J. H. 1996. Morphology and ultrastructure of possible integumentary sense organs in the estuarine crocodile (*Crocodylus porosus*). *Journal of Morphology* 229:315–24.

10. SOARES, D. 2002. An ancient sensory organ in crocodilians. *Nature* 417:241–2.

11. LEITCH, D. B., and CATANIA, K. C. 2012. Structure, innervation and response properties of integumentary sensory organs in crocodilians. *Journal of Experimental Biology* 215:4217–30.

12. DI-POÏ, N., and MILINKOVITCH, M. C. 2013. Crocodylians evolved scattered multi-sensory micro-organs. *EvoDevo* 4:19.

13. CARR, T. D., VARRICCHIO, D. J., SEDLMAYR, J. C., ROBERTS, E. M., and MOORE, J. R. 2017. A new tyrannosaur with evidence for anagenesis and crocodile-like facial sensory system. *Scientific Reports* 7:44942.

14. KLÄRNER, D., and BARTH, F. G. 1982. Vibratory signals and prey capture in orb-weaving spiders (*Zygiella x-notata, Nephila clavipes*; Araneidae). *Journal of Comparative Physiology A* 148:445–55.

15. BARTH, F. G. 2002. Spider senses—technical perfection and biology. *Zoology* 105:271–85.

16. BARTH, F. G., and GEETHABALI. 1982. Spider vibration receptors: threshold curves of individual slits in the metatarsal Lyriform organ. *Journal of Comparative Physiology A* 148:175–85.

17. BOY, C. V. 1880. The influence of a tuning-fork on the garden spider. *Nature* 23:149–50.

18. MASTERS, W. M., and MARKL, H. 1981. Vibration signal transmission in spider orb webs. *Science* 213:363–5.

19. BAURECHT, D., and BARTH, F. G. 1992. Vibratory communication in spiders. *Journal of Comparative Physiology A* 171:231–43.

CHAPTER 6

1. CAPRIO, J., SHIMOHARA, M., MARUI, T., HARADA, S., and KIYOHARA, S. 2014. Marine teleost locates live prey through pH sensing. *Science* 344:1154–6.

2. MASSE, N. Y., TURNER, G. C., and JEFFERIS, G. S. X. E. 2009. Olfactory information processing in *Drosophila*. *Current Biology* 19: R700–R713.

3. SASS, H. 1983. Production, release and effectiveness of two female sex pheromone components of *Periplaneta americana*. *Journal of Comparative Physiology A* 152:309–17.

4. HÖLLDOBLER, B., and WILSON, E. O. 1994. *Journey to the Ants: A Story of Scientific Exploration*. Cambridge, MA and London: The Belknap Press of Harvard University Press.

5. A distinction in olfaction is often made between so-called signature mixes, which differ between individuals and groups and are used for recognition, and pheromones. The former often need to be learnt and vary among individuals/groups, whereas the latter are usually species-specific and produce stereotyped responses. See:
   WYATT, T. D. 2014. *Pheromones and Animal Behavior: Chemical Signals and Signatures*, 2nd edn. Cambridge: Cambridge University Press.
   WYATT, T. D. 2017. Pheromones. *Current Biology* 27:R739–R743.

6. MONNIN, T., RATNIEKS, F. L. W., JONES, G. R., and BEARD, R. 2002. Pretender punishment induced by chemical signalling in a queenless ant. *Nature* 419:61–3.

7. YAN, H., et al. 2017. An engineered orco mutation produces aberrant social behavior and defective neural development in ants. *Cell* 170:736–47.

8. OZAKI, M., WADA-KATSUMATA, A., FUJIKAWA, K., IWASAKI, M., YOKOHARI, F., SATAOJI, Y., NISIMURA, T., and YAMAOKA, R. 2005. Ant nestmate and non-nestmate discrimination by a chemosensory sensillum. *Science* 309:311–14.

9. SHARMA, K. R., ENZMANN, B. L., SCHMIDT, Y., MOORE, D., JONES, G. R., PARKER, J., BERGER, S. L., REINBERG, D., ZWIEBEL, L. J., BREIT, B., LIEBIG, J., and RAY, A. 2015. Cuticular hydrocarbon pheromones for social behavior and their coding in the ant antenna. *Cell Reports* 12:1261–71.

D'ETTORRE, P., DEISIG, N., and SANDOZ, J.-C. 2017. Decoding ants' olfactory system sheds light on the evolution of social communication. *PNAS* 114:8911–13.

PASK, G. M., SLONE, J. D., MILLAR, J. G., DAS, P., MOREIRA, J. A., ZHOU, X., BELLO, J., BERGER, S. L., BONASIO, R., DESPLAN, C., REINBERG, D., LIEBIG, J., ZWIEBEL, L. J., and RAY, A. 2017. Specialized odorant receptors in social insects that detect cuticular hydrocarbon cues and candidate pheromones. *Nature Communications* 8:297.

SLONE, J. D., PASK, G. M., FERGUSON, S. T., MILLAR, J. G., BERGER, S. L., REINBERG, D., LIEBIG, J., RAY, A., and ZWIEBEL, L. J. 2017. Functional characterization of odorant receptors in the ponerine ant, *Harpegnathos saltator. PNAS* 114:8586–91.

10. BUCK, L., and AXEL, R. 1991. A novel multigene family may encode odorant receptors: a molecular basis for odor recognition. *Cell* 65:175–87.

11. QUIGNON, P., RIMBAULT, M., ROBIN, S., and GALIBERT, F. 2012. Genetics of canine olfaction and receptor diversity. *Mammal Genome* 23:132–43.

GALIBERT, F., AZZOUZI, N., QUIGNON, P., and CHAUDIEU, G. 2016. The genetics of canine olfaction. *Journal of Veterinary Behavior* 16:86–93.

12. CRAVEN, B. A., PATERSON, E. G., and SETTLES, G. S. 2010. The fluid dynamics of canine olfaction: unique nasal airflow patterns as an explanation of macrosmia. *Journal of the Royal Society Interface* 7:933–43.

13. CATANIA, K. C. 2013. Stereo and serial sniffing guide navigation to an odour source in a mammal. *Nature Communications* 4:1441–8.

CHAPTER 7

1. Fijn, R. C., Hiemstra, D., Phillips, R. A., and van der Winden, J. 2013. Arctic terns *Sterna paradisaea* from the Netherlands migrate record distances across three oceans to Wilkes Land, East Antarctica. *Ardea* 101:3–12.

2. Chernetsov, N., Kishkinev, D., and Mouritsen, H. 2008. A long-distance avian migrant compensates for longitudinal displacement during spring migration. *Current Biology* 18:188–90.

3. Lohmann, K. J. 1991. Magnetic orientation by hatchling loggerhead sea turtles (*Caretta caretta*). *Journal of Experimental Biology* 155:37–49.

4. Lohmann, K. J., and Lohmann, C. M. F. 1994. Detection of magnetic inclination angle by sea turtles: a possible mechanism for determining latitude. *Journal of Experimental Biology* 194:23–32.

   Lohmann, K. J., and Lohmann, C. M. F. 1996. Detection of magnetic field intensity by sea turtles. *Nature* 380:59–61.

   Lohmann, K. J., Cain, S. D., Dodge, S. A., and Lohmann, C. M. F. 2001. Regional magnetic fields as navigational markers for sea turtles. *Science* 294:364–6.

   Putman, N. F., Endres, C. S., Lohmann, C. M. F., and Lohmann, K. J. 2011. Longitude perception and bicoordinate magnetic maps in sea turtles. *Current Biology* 21:463–6.

5. Lohmann, K. J., Lohmann, C. M. F., Ehrhart, L. M., Bagley, D. A., and Swing, T. 2004. Geomagnetic map used in sea-turtle navigation. *Nature* 428:909–10.

6. Fuxjager, M. J., Davidoff, K. R., Mangiamele, L. A., and Lohmann, K. J. 2015. The geomagnetic environment in which sea turtle eggs incubate affects subsequent magnetic navigation behaviour of hatchlings. *Proceedings of the Royal Society of London.* Series B 281:20141218.

7. Brothers, J. R., and Lohmann, K. J. 2015. Evidence for geomagnetic imprinting and magnetic navigation in the natal homing of sea turtles. *Current Biology* 25:392–6.

   Brothers, J. R., and Lohmann, K. J. 2018. Evidence that magnetic navigation and geomagnetic imprinting shape spatial genetic variation in sea turtles. *Current Biology* 28:1325–9.

8. JAMES, M. C., MYERS, R. A., and OTTENSMEYER, A. 2005. Behaviour of leatherback sea turtles, *Dermochelys coriacea*, during the migratory cycle. *Proceedings of the Royal Society of London. Series B* 272:1547–55.

9. QUINN, T. P. 1980. Evidence for celestial and magnetic compass orientation in lake migrating sockeye salmon fry. *Journal of Comparative Physiology A* 137:243–8.

QUINN, T. P., and BRANNON, E. L. 1982. The use of celestial and magnetic cues by orienting sockeye salmon smolts. *Journal of Comparative Physiology A* 147:547–52.

10. PUTMAN, N. F., LOHMANN, K. J. PUTMAN, E. M., QUINN, T. P., KLIMLEY, A. P., and NOAKES, D. L. G. 2013. Evidence for geomagnetic imprinting as a homing mechanism in pacific salmon. *Current Biology* 23:312–16.

PUTMAN, N. F., JENKINS, E. S., MICHIELSENS, C. G. J., and NOAKES, D. L. G. 2014. Geomagnetic imprinting predicts spatiotemporal variation in homing migration of pink and sockeye salmon. *Journal of the Royal Society Interface* 11:20140542.

11. MANN, S., SPARKS, N. H. C., WALKER, M. M., and KIRSCHVINK, J. L. 1988. Ultrastructure, morphology and organization of biogenic magnetite from sockeye salmon, *Oncorhynchus nerka*: implications for magnetoreception. *Journal of Experimental Biology* 140:35–49.

WALKER, M. M., QUINN, T. P., KIRSCHVINK, J. L., and GROOT, C. 1988. Production of single-domain magnetite throughout life by sockeye salmon *Oncorhynchus nerka*. *Journal of Experimental Biology* 140:51–63.

12. CHEW, G. L., and BROWN, G. E. 1989. Orientation of rainbow trout (*Salmo gairdneri*) in normal and null magnetic fields. *Canadian Journal of Zoology* 67:641–3.

HELLINGER, J., and HOFFMANN, K.-P. 2009. Magnetic field perception in the rainbow trout, *Oncorhynchus mykiss*. *Journal of Comparative Physiology A* 195:873–9.

PUTMAN, N. F., MEINKE, A. M., and NOAKES, D. L. G. 2014. Rearing in a distorted magnetic field disrupts the 'map sense' of juvenile steelhead trout. *Biology Letters* 10:20140169.

WALKER, M. M., DIEBEL, C. E., HAUGH, C. V., PANKHURST, P. M., MONTGOMERY, J. C., and GREEN, C. R. 1997. Structure and function of the vertebrate magnetic sense. *Nature* 390:371–6.

13. HELLINGER, J., and HOFFMANN, K.-P. 2012. Magnetic field perception in the rainbow trout *Oncorynchus mykiss*: magnetite mediated, light dependent or both? *Journal of Comparative Physiology A* 198:593–605.

14. FITAK, R. R., WHEELER, B. R., ERNST, D. A., LOHMANN, K. J., and JOHNSEN, S. 2017. Candidate genes mediating magnetoreception in rainbow trout (*Oncorhynchus mykiss*). *Biology Letters* 13:20170142.

   FITAK, R. R., SCHWEIKERT, L. E., WHEELER, B. R., ERNST, D. A., LOHMANN, K. J., and JOHNSEN, S. 2018. Near absence of differential gene expression in the retina of rainbow trout after exposure to a magnetic pulse: implications for magnetoreception. *Biology Letters* 14:20180209.

15. WILTSCHKO, W., and WILTSCHKO, R. 1996. Magnetic orientation in birds. *Journal of Experimental Biology* 199:29–38.

   WILTSCHKO, W., and WILTSCHKO, R. 1972. Magnetic compass of European robins. *Science* 176:62–4.

16. WILTSCHKO, W. 1974. Evidence for an innate magnetic compass in garden warblers. *Naturwissenschaften* 61:406–7.

17. KEETON, W. T. 1971. Magnets interfere with pigeon homing. *PNAS* 68:102–6.

18. BEASON, R. C., and NICHOLS, J. E. 1984. Magnetic orientation and magnetically sensitive material in a transequatorial migratory bird. *Nature* 309:151–3.

19. WILTSCHKO, W., MUNRO, U., BEASON, R. C., FORD, H., and WILTSCHKO, R. 1994. A magnetic pulse leads to a temporary deflection in the orientation of migratory bird. *Experientia* 50:697–700.

20. WILTSCHKO, W., MUNRO, U., FORD, H., and WILTSCHKO, R. 1993. Red light disrupts magnetic orientation of migratory birds. *Nature* 364:525–7.

   WILTSCHKO, W., and WILTSCHKO, R. 1995. Migratory orientation of European robins is affected by the wavelength of light as well as by a magnetic pulse. *Journal of Comparative Physiology A* 177:363–9.

21. MOURITSEN, H., JANSSEN-BIENHOLD, U., LIEDVOGEL, M., FEENDERS, G., STALLEICKEN, J., DIRKS, P., and WEILER, R. 2004. Cryptochromes and neuronal-activity markers colocalize in the retina of migratory birds during magnetic orientation. *PNAS* 101:14294–9.

22. NIESSNER, C., DENZAU, S., GROSS, J. C., PEICHL, L., BISCHOF, H.-J., FLEISSNER, G., WILTSCHKO, W., and WILTSCHKO, R. 2011. Avian ultraviolet/violet cones identified as probable magnetoreceptors. *PLoS ONE* 6:e20091.

23. MOURITSEN, H., FEENDERS, G., LIEDVOGEL, M., WADA, K., and JARVIS, E. D. 2005. Night-vision brain area in migratory songbirds. *PNAS* 102:8339–44.

  HEYERS, D., MANNS, M., LUKSCH, H., GÜNTÜRKÜN, O., and MOURITSEN, H. 2007. A visual pathway links brain structures active during magnetic compass orientation in migratory birds. *PLoS ONE* 2:e937.

  ZAPKA, M., HEYERS, D., HEIN, C. M., ENGELS, S., SCHNEIDER, N.-L., HANS, J., WEILER, S., DREYER, D., KISHKINEV, D., WILD, J. M., and MOURITSEN, H. 2009. Visual but not trigeminal mediation of magnetic compass information in a migratory bird. *Nature* 461:1274–8.

24. NORDMANN, G. C., HOCHSTOEGER, T., and KEAYS, D. A. 2017. Magnetoreception—a sense without a receptor. *PLoS Biology* 15:e2003234.

25. BEGALL, S., ČERVENÝ, J., NEEF, J., VOJTECH, O., and BURDA, H. 2008. Magnetic alignment in grazing and resting cattle and deer. *PNAS* 105:13451–5.

  BEGALL, S., MALKEMPER, E. P., ČERVENÝ, J., NĚMEC, P., and BURDA, H. 2013. Magnetic alignment in mammals and other animals. *Mammalian Biology* 78:10–20.

  WEIJERS, D., HEMERIK, L., and HEITKÖNIG, I. M. A. 2018. An experimental approach in revisiting the magnetic orientation of cattle. *PLoS ONE* 13:e0187848.

26. HART, V., NOVÁKOVÁ, P., MALKEMPER, E. P., BEGALL, S., HANZAL, V., JEŽEK, M., KUŠTA, T., NĚMCOVÁ, V., ADÁMKOVÁ, J., BENEDIKTOVÁ, K., ČERVENÝ, J., and BURDA, H. 2013. Dogs are sensitive to small variations of the Earth's magnetic field. *Frontiers in Zoology* 10:80.

CHAPTER 8

1. POLING, A., WEETJENS, B., COX, C., BEYENE, N., BACH, H., and SULLY, A. 2010. Teaching giant African pouched rats to find landmines:

operant conditioning with real consequences. *Behaviour Analysis in Practice* 3:19–25.

EDWARDS, T. L., COX, C., WEETJENS, B., and TEWELDE, T. 2015. Giant African pouched rats (*Cricetomys gambianus*) that work on tilled soil accurately detect land mines. *Journal of Applied Behavior Analysis* 48:696–700.

POLING, A., WEETJENS, B., COX, C., BEYENE, N. W., BACH, H., and SULLY, A. 2011. Using trained pouched rats to detect land mines: another victory for operant conditioning. *Journal of Applied Behavior Analysis* 44:351–5.

2. KANHERE, E., WANG, N., ASADNIA, M., KOTTAPALLI, A. G. P., and MIAO, J. M. 2015. Crocodile inspired dome pressure sensor for hydrodynamic sensing. *Solid-State Sensors, Actuators and Microsystems (TRANSDUCERS)*. Alaska, AK: IEEE Xplore, 1199–202.

KANHERE, E., WANG, N., KOTTAPALLI, A. G. P., ASADNIA, M., SUBRAMANIAM, V., MIAO, J., and TRIANTAFYLLOU, M. 2016. Crocodile-inspired dome-shaped pressure receptors for passive hydrodynamic sensing. *Bioinspiration & Biomimetics* 11:056007.

KANHERE, E., WANG, N., KOTTAPALLI, A. G. P., SUBRAMANIAM, V., MIAO, J. M., and TRIANTAFYLLOU, M. S. 2017. Crocodile-inspired dome shaped sensors for underwater object detection. *IEEE Sensors* 1–3.

3. FISHMAN, A., ROSSITER, J., and HOMER, M. 2015. Hiding the squid: patterns in artificial cephalopod skin. *Journal of the Royal Society Interface* 12:20150281.

4. GREENWOOD, V. J., SMITH, E. L., GOLDSMITH, A. R., CUTHILL, I. C., CRISP, L. H., WALTER-SWAN, M. B., and BENNETT, A. T. D. 2004. Does the flicker frequency of fluorescent lighting affect the welfare of captive European starlings? *Applied Animal Behaviour Science* 86:145–59.

EVANS, J. E., CUTHILL, I. C., and BENNETT, T. D. 2006. The effect of flicker from fluorescent lights on mate choice in captive birds. *Animal Behaviour* 72:393–400.

EVANS, J. E., SMITH, E. L., BENNETT, A. T. D., CUTHILL, I. C., and BUCHANAN, L. 2012. Short-term physiological and behavioural effects of high- versus low-frequency fluorescent light on captive birds. *Animal Behaviour* 83:25–33.

SMITH, E. L., EVANS, J. E., and PÁRRAGA, C. A. 2005. Myoclonus induced by cathode ray tube screens and low-frequency lighting in the European starling (*Sturnus vulgaris*). *The Veterinary Record* 157:148–50.

5. INGER, R., J., DAVIES, T. W., and GASTON, J. 2014. Potential biological and ecological effects of flickering artificial light. *PLoS ONE* 9:e98631.

6. PAUL, S. C., and STEVENS, M. 2020. Horse vision and obstacle visibility in horseracing. *Applied Animal Behaviour Science* 222:104882.

7. WOLF, M. C., and MOORE, P. A. 2002. Effects of the herbicide metolachlor on the perception of chemical stimuli by *Orconectes rusticus*. *Journal of the North American Benthological Society* 21:457–67.

COOK, M. E., and MOORE, P. A. 2008. The effects of the herbicide metolachlor on agonistic behavior in the crayfish, *Orconectes rusticus*. *Archives of Environmental Contamination and Toxicology* 55:94–102.

LAHMAN, S. E., TRENT, K. R., and MOORE, P. A. 2015. Sublethal copper toxicity impairs chemical orientation in the crayfish, *Orconectes rusticus*. *Ecotoxicology and Environmental Safety* 113:369–77.

8. FISHER, H. S., WONG, B. B. M., and ROSENTHAL, G. G. 2006. Alteration of the chemical environment disrupts communication in a freshwater fish. *Proceedings of the Royal Society of London. Series B* 273:1187–93.

9. NEVITT, G. A., VEIT, R. R., and KAREIVA, P. 1995. Dimethyl sulphide as a foraging cue for Antarctic Procellariiform seabirds. *Nature* 376:680–2.

DELL'ARICCIA, G., CÉLÉRIER, A., GABIROT, M., PALMAS, P., MASSA, B., and BONADONNA, F. 2014. Olfactory foraging in temperate waters: sensitivity to dimethylsulphide of shearwaters in the Atlantic Ocean and Mediterranean Sea. *Journal of Experimental Biology* 217:1701–9.

NEVITT, G. A., and BONADONNA, F. 2005. Sensitivity to dimethyl sulphide suggests a mechanism for olfactory navigation by seabirds. *Biology Letters* 1:303–5.

SAVOCA, M. S., and NEVITT, G. A. 2014. Evidence that dimethyl sulfide facilitates a tritrophic mutualism between marine primary producers and top predators. *PNAS* 111:4157–61.

10. SAVOCA, M. S., WOHLFEIL, M. E., EBELER, S. E., and NEVITT, G. A. 2016. Marine plastic debris emits a keystone infochemical for olfactory foraging seabirds. *Science Advances* 2:e1600395.

11. SLABBEKOORN, H., and PEET, M. 2003. Birds sing at a higher pitch in urban noise. *Nature* 424:267.
    SLABBEKOORN, H., and BOER-VISSER, A. DEN. 2006. Cities change the songs of birds. *Current Biology* 16:2326–31.

12. FOOTE, A. D., OSBORNE, R. W., and HOELZEL, A. R. 2004. Whale-call response to masking boat noise. *Nature* 428:910.
    MILLER, P. J. O., BIASSONI, N., SAMUELS, A., and TYACK, P. L. 2000. Whale songs lengthen in response to sonar. *Nature* 405:903–4.

13. WALE, M. A., SIMPSON, S. D., and RADFORD, A. N. 2013. Noise negatively affects foraging and antipredator behaviour in shore crabs. *Animal Behaviour* 86:111–18.
    CARTER, E. E., TREGENZA, T., and STEVENS, M. 2020. Ship noise inhibits colour change, camouflage, and anti-predator behaviour in shore crabs. *Current Biology* 30:R211–R212.
    WALE, M. A., SIMPSON, S. D., and RADFORD, A. N. 2013. Size-dependent physiological responses of shore crabs to single and repeated playback of ship noise. *Biology Letters* 9:20121194.

14. PHILLIPS, J. N., RUEF, S. K., GARVIN, C. M., LE, M-Y. T., and FRANCIS, C. D. 2019. Background noise disrupts host-parasitoid interactions. *Royal Society Open Science* 6:190867.

15. VAN LANGEVELDE, F., ETTEMA, J. A., DONNERS, M., WALLISDEVRIES, M. F., and GROENENDIJK, D. 2011. Effect of spectral composition of artificial light on the attraction of moths. *Biological Conservation* 144:2274–81.

16. KNOP, E., ZOLLER, L., RYSER, R., GERPE, C., HÖRLER, M., and FONTAINE, C. 2017. Artificial light at night as a new threat to pollination. *Nature.* 548:206–9.
    MACGREGOR, C. J., EVANS, D. M., FOX, R., and POCOCK, M. J. O. 2017. The dark side of street lighting: impacts on moths and evidence for the disruption of nocturnal pollen transport. *Global Change Biology* 23:697–707.

17. ALTERMATT, F., and EBERT, D. 2016. Reduced flight-to-light behaviour of moth populations exposed to long-term urban light pollution. *Biology Letters* 12:20160111.

18. HEILING, A. M. 1999. Why do nocturnal orb-web spiders (*Araneidae*) search for light? *Behavioral Ecology and Sociobiology* 46:43–9.

WILLMOTT, N. J., HENNEKEN, J., ELGAR, M. A., and JONES, T. M. 2019. Guiding lights: foraging responses of juvenile nocturnal orb-web spiders to the presence of artificial light at night. *Ethology* 125:289–97.

19. BURDA, H., BEGALL, S., ČERVENÝ, J., NEEF, J., and NĚMEC, P. 2009. Extremely low-frequency electromagnetic fields disrupt magnetic alignment of ruminants. *PNAS* 106:5708–13.

20. ENGELS, S., SCHNEIDER, N.-L., LEFELDT, N., HEIN, C. M., ZAPKA, M., MICHALIK, A., ELBERS, D., KITTEL, A., HORE, P. J., and MOURITSEN, H. 2014. Anthropogenic electromagnetic noise disrupts magnetic compass orientation in a migratory bird. *Nature* 509:353–61.

21. MUNDAY, P. L., DIXSON, D. L., DONELSON, J. M., JONES, G. P., PRATCHETT, M. S., DEVITSINA, G. V., and DØVING, K. B. 2009. Ocean acidification impairs olfactory discrimination and homing ability of a marine fish. *PNAS* 106:1848–52.

CLARK, T. D., RABY, G. D., ROCHE, D. G., BINNING, S. A., SPEERS-ROESCH, B., JUTFELT, F., and SUNDIN, J. 2020. Ocean acidification does not impair the behaviour of coral reef fish. *Nature* 577:370–5.

DIXON, D. L., MUNDAY, P. L., and JONES, G. P. 2010. Ocean acidification disrupts the innate ability of fish to detect predator olfactory cues. *Ecology Letters* 13:68–75.

LEDUC, A. O. H. C., MUNDAY, P. L., BROWN, G. E., and FERRARI, M. C. O. 2013. Effects of acidification on olfactorymediated behaviour in freshwater and marine ecosystems: a synthesis. *Philosophical Transactions of the Royal Society B* 368:20120447.

MUNDAY, P. L., DIXON, D. L., McCORMICK, M. I., MEEKAN, M., FERRARI, M. C. O., and CHIVERS, P. 2010. Replenishment of fish populations is threatened by ocean acidification. *PNAS* 107:12930–4.

PORTEUS, C. S., HUBBARD, P. C., WEBSTER, T. M. U., AERLE, R. van, CANÁRIO, A. V. M., SANTOS, E. M., and WILSON, W. 2018. Near-future CO2 levels impair the olfactory system of a marine fish. *Nature Climate Change* 8:737–43.

22. GORDON, T. A. C., RADFORD, A. N., DAVIDSON, I. K., BARNES, K., McCLOSKEY, K., NEDELEC, S. L., MEEKAN, M. G., McCORMICK, M. I., and SIMPSON, S. D. 2019. Acoustic enrichment can enhance fish community development on degraded coral reef habitat. *Nature Communications* 10:5414.

23. MILLS, L. S., ZIMOVA, M., OYLER, J., RUNNING, S., ABATZOGLOU, J. T., and LUKACS, P. M. 2013. Camouflage mismatch in seasonal coat color due to decreased snow duration. *PNAS* 110:7360–5.

   ZIMOVA, M., MILLS, L. S., LUKACS, P. M., and MITCHELL, M. S. 2014. Snowshoe hares display limited phenotypic plasticity to mismatch in seasonal camouflage. *Proceedings of the Royal Society of London*. Series B 281:20140029.

   ZIMOVA, M., MILLS, L. S., and NOWAK, J. J. 2016. High fitness costs of climate change-induced camouflage mismatch. *Ecology Letters* 19:299–307.

24. WANG, J. H., BOLES, L. C., HIGGINS, B., and LOHMANN, K. J. 2007. Behavioral responses of sea turtles to lightsticks used in longline fisheries. *Animal Conservation* 10:176–82.

   GLESS, J. M., SALMON, M., and WYNEKEN, J. 2008. Behavioral responses of juvenile leatherbacks *Dermochelys coriacea* to lights used in the longline fishery. *Endangered Species Research* 5:239–47.

   ORTIZ, N., MANGEL, J. C., WANG, J., ALFARO-SHIGUETO, J., PINGO, S., JIMENEZ, A., SUAREZ, T., SWIMMER, Y., CARVALHO, F., and GODLEY, B. J. 2016. Reducing green turtle bycatch in small-scale fisheries using illuminated gillnets: the cost of saving a sea turtle. *Marine Ecology Progress Series* 545:251–9.

   WANG, J., BARKAN, J., FISLER, S., GODINEZ-REYES, C., and SWIMMER, Y. 2013. Developing ultraviolet illumination of gillnets as a method to reduce sea turtle bycatch. *Biology Letters* 9:20130383.

   WANG, J. H., BOLES, L. C., HIGGINS, B., and LOHMANN, K. J. 2007. Behavioral responses of sea turtles to lightsticks used in longline fisheries. *Animal Conservation* 10:176–82.

   WANG, J. H., SHARA, F., and SWIMMER, Y. 2010. Developing visual deterrents to reduce sea turtle bycatch in gill net fisheries. *Marine Ecology Progress Series* 408:241–50.

   WANG, J., BARKAN, J., FISLER, S., GODINEZ-REYES, C., and SWIMMER, Y. 2013. Developing ultraviolet illumination of gillnets as a method to reduce sea turtle bycatch. *Biology Letters* 9:20130383.

25. VEITS, M., et al. 2019. Flowers respond to pollinator sound within minutes by increasing nectar sugar concentration. *Ecology Letters* 22:1483–92.

LIST OF IMAGE PERMISSIONS

Figure 24. Neil Bromhall / Shutterstock

Figure 25. © Duncan B. Leitch

Figure 26. Courtesy of Arthropod Structure & Development 47.2

Figure 27. Megan McCarty and Jason Sturner

Figure 28. A. Craven Brent, Eric G. Paterson, and Gary S. Settles (2010), 'The fluid dynamics of canine olfaction: unique nasal airflow patterns as an explanation of macrosmia', *J. R. Soc. Interface* 7933–43

Figure 30. L. Shyamal / Wikipedia

Figure 31. R. Wiltschko and W. Wiltschko (2014), 'Sensing magnetic directions in birds: radical pair processes involving cryptochrome', *Biosensors* 4: 221–42. Figure 4B

Figure 32. Aolesak / Wikipedia

Figure 33. NASA Earth Observatory

Plate 1. Martin Stevens / Sensory Ecology Lab

Plate 2. Courtesy Matthew L.M. Lim

Plate 3. Courtesy Matthew L.M. Lim

Plate 4. C. Taboada, A.E. Brunetti, F.N. Pedron, F.C. Neto, D.A. Estrin, S.E. Bari, L.B. Chemes, N.P. Lopes, M.G. Lagorio, and J. Faivovich (2017), 'Naturally occurring fluorescence in frogs' *PNAS* 114: 3672–7. Figure 1A

Plate 5. Martin Stevens

Plate 7. Charles J. Sharp

Plate 8. Dr. Liqun Luo, Stanford University

Plate 9. Sarah Paul and Martin Stevens

Plate 10. Emily Carter and Martin Stevens

Plate 11. ProDelphinus

INDEX